25p

THE SCHOOL MATHEMATICS PROJECT

When the S.M.P. was founded in 1961, its objective was to devise radically new mathematics courses, with accompanying G.C.E. syllabuses and examinations, which would reflect, more adequately than did the traditional syllabuses, the up-to-date nature and usages of mathematics.

The first stage of this objective is now more or less complete. *Books 1–5* form the main series of pupils' texts, starting at the age of 11+ and leading to the O-level examination 'S.M.P. Mathematics', while *Books 3T, 4* and *5* give a three-year course to the same O-level examination. (*Books T* and *T4*, together with their Supplement, represent the first attempt at this three-year course, but they may be regarded as obsolete.) *Advanced Mathematics Books 1–4* cover the syllabus for the A-level examination in 'S.M.P. Mathematics' and in preparation are five (or more) shorter texts covering the material of various sections of the A-level examination in 'S.M.P. Further Mathematics'. There are two books for 'S.M.P. Additional Mathematics' at O-level. Every book is accompanied by a Teacher's Guide.

For the convenience of schools, the S.M.P. has an arrangement whereby its examinations are made available by every G.C.E. Examining Board, and it is most grateful to the Secretaries of the eight Boards for their cooperation in this. At the same time, most Boards now offer their own syllabuses in 'modern mathematics' for which the S.M.P. texts are suitable.

By 1967, it had become clear from experience in comprehensive schools that the mathematical content of the S.M.P. texts was suitable for a much wider range of pupil than had been originally anticipated, but that the presentation needed adaptation. Thus it was decided to produce a new series, *Books A–H*, which could serve as a secondary-school course starting at the age of 11+. These books are specially suitable for pupils aiming at a C.S.E. examination; however, the framework of the C.S.E. examinations is such that it is inappropriate for the S.M.P. to offer its own examination as it does for the G.C.E.

The completion of all these books does not mean that the S.M.P. has no more to offer to the cause of curriculum research. The team of S.M.P. writers, now numbering some thirty school and university mathematicians, is continually testing and revising old work and preparing for new. At the same time, the effectiveness of the S.M.P.'s work depends, as it always has done, on obtaining reactions from active teachers—and also from pupils—in the classroom. Readers of the texts can therefore send their comments to the S.M.P. in the knowledge that they will be warmly welcomed.

Finally, the year-by-year activity of the S.M.P. is recorded in the annual Director's reports which readers are enc[illegible] request to the S.M.P. Office at Westfield Colleg[illegible] Kidderpore Avenue, Hampstead, London NW3 [illegible]

ACKNOWLEDGEMENTS

The principal authors, on whose contributions the S.M.P. texts are largely based, are named in the annual Reports. Many other authors have also provided original material, and still more have been directly involved in the revision of draft versions of chapters and books. The Project gratefully acknowledges the contributions which they and their schools have made.

This book—*Teacher's Guide for Book A*—has been prepared by

Catherine Braithwaite
D. Dorrian
Joyce Harris
K. Lewis
W. Mrozowski
Margaret Wilkinson
E. Williamson

and edited by P. G. Bowie assisted by Elizabeth Evans.

The Project is most grateful for the advice on points of fundamental mathematics given by Dr J. V. Armitage.

The drawings in this book are by Penny Wager.

The Project is grateful to the BBC for permission to use the photograph of Harry Worth and the sketch of a dalek.

We are much indebted to the Cambridge University Press for their cooperation and help at all times in the preparation of this book.

The Project owes a great deal to its Secretary, Miss A. J. Freeman; also to Mrs E. L. Humphreys and Mrs E. Muir for their assistance and for their typing in connection with this book.

THE SCHOOL MATHEMATICS PROJECT

TEACHER'S GUIDE FOR BOOK A [METRIC]

CAMBRIDGE
AT THE UNIVERSITY PRESS

1971

Published by the Syndics of the Cambridge University Press
Bentley House, 200 Euston Road, London NW1 2DB
American Branch: 32 East 57th Street, New York, N.Y.10022

Library of Congress Catalogue Card Number: 68–21399
ISBN: 0 521 07839 3

First published 1968
Metric edition 1970
Reprinted 1971

Printed in Great Britain
at the University Printing House, Cambridge
(Brooke Crutchley, University Printer)

Preface—Book A

This is the first of eight books designed to cover a course suitable for those who wish to take a CSE Examination on one of the reformed mathematics syllabuses.

The material is based upon the first four books of the O-level series, SMP Books 1–4. The connection is maintained to the extent that it will be possible to change from one series to the other at the end of the first year, or even at a later stage. For example, having started with Books A and B, a pupil will be able to move to Book 2.

Within each year's work, the material has been entirely broken down and rewritten. The differences between the two series fall under five headings:

(i) The emphasis placed upon preliminary investigation and question in the O-level books has been taken up and developed in this series. From the Prelude, which is entirely experimental, through to the chapter on polyhedra, which involves a very great deal of practical work, each chapter involves activity and investigation as well as establishing the main points by a process of question and answer.

(ii) The O-level books have tended to assume rather more knowledge of some subjects than has been found entirely justified for the average pupil. In this new series, more time is given to the earlier stages and some sections on preliminary work have been added.

(iii) The original chapters have been divided into several parts for two reasons: first, to allow pupils more time to absorb one point before moving on to its development; secondly, by providing a smaller interval between one discussion of a subject and the next, to reduce the need for extensive revision.

(iv) Preludes and Interludes have been added. In general, the Preludes will be an integral part of the book. Later chapters will often depend upon, and sometimes specifically refer to, the experiences gained from them. The Interludes will tend to be separate from the main development of the text. In general, they will describe situations from which mathematics can be drawn. Most of these situations will be internal to the classroom, as in the designing of patterns described in Book A; but some, such as the surveying in Book C, will involve work outside the classroom. In all cases, pupils should be encouraged to formulate their own questions, and so to realize contexts within which mathematics is developed.

(v) There have been some small changes in mathematical content taken from Book 1 and similar small changes may be expected in later books. For example, there is no chapter on Sets. The subject and the notation have been introduced where necessary in Book A, but the study of Venn diagrams and of their attendant problems has been postponed to a later book. Some statistics and topology have been brought into Book B; also a consideration of non-numerical relations. These three subjects, all fundamental to the course, are started in Book 2, but they are interesting and suitable for Book B.

The main differences between the content of these two SMP series and that of the more traditional texts arise from two convictions: first, that understanding and interest in the general statements of mathematics stem from experience of a wide range of particular situations and from confidence that questions of mathematical significance can be asked and answered in many of these situations; secondly, that these experiences should arise inside as well as outside mathematics. Thus, it is useful to have discussed and manipulated expressions in symbols concerned with sets, transformations and matrices in order to obtain a deeper understanding of the significance of these symbols, before perfecting the techniques of equation-solving within the field of real numbers. It is worth discovering a wide variety of relations and of methods for illustrating them, before applying this knowledge to a discussion of mathematical functions, equations and graphs. It is useful to gain considerable experience of shapes and of methods of measuring and describing them by studying polyhedra and by considering transformations of figures in the plane, before developing any formal body of geometrical theorems.

For these reasons, the newer topics should be considered to be an integral part of the course, not something to be learnt at the side of, and separate from it.

In this book, Book A, two chapters are devoted to consideration of the patterns among the counting numbers; these lead to sections on various sequences of numbers as well as on factors and multiples. Two short chapters are included to remind pupils of the basic ideas, but not the manipulative techniques, of fractions; a chapter on number bases emphasizes the importance of position in the numeral system, and this prepares for the chapter on decimals. The chapter on coordinates lays a foundation for later work on graphs. The several geometry chapters—on angle, symmetry, polygons and polyhedra—are all designed to give experience of shape and methods of describing it, though the problems of measurement of length have been delayed to Book B.

Preface—Teacher's Guide for Book A

This book contains the pages from pupil's Book A interleaved with comments and answers to questions. The comments have three aims in mind. They try to show the purpose of new and unusual topics and fresh approaches to old topics; they provide occasional notes and references on the background to the mathematics and they discuss how the contents of Book A can best be presented to classes.

None of the new material in the pupil's book was included until it had been tried in classes and found to succeed. Much that is relatively unfamiliar, for example the work on networks and statistics in Book B, is particularly relevant to the present-day world. Sets and functions provide a deeper understanding of some familiar elements in the course and reduce the need for tedious and time-consuming drill. In other cases, for example with transformation geometry, the new approach is more suitable for younger pupils.

In the pupil's book we tried to produce a mathematics text which pupils could use with only occasional help from their teachers. We hope that the books will initiate activities and investigations beyond the immediate problem, and sometimes outside the classroom itself, while indicating the lines of development to which the pupils can eventually return.

The background notes give references to those sources which we feel are of the greatest direct value, though there are many books of more general interest. We hope that teachers who have used this guide in conjunction with the pupil's book will send us comments and criticisms. Only in this way can we be sure that the guide will continue to serve its purpose.

Publisher's note

The book consists of the pupil's Book A interleaved with pages of answers and commentary for the teacher; the teacher's pages are distinguished by a red line down the margin.

The book is numbered in red throughout with the prefix T, e.g. T39. For ease of reference pupil's pages also retain the black number that appears on the identical page in Book A. The teacher's numbering is used in the list of contents and the index, and the list of contents also repeats the pupil's number for the beginning of each chapter.

Contents

Contents

Equipment and materials

Lead and coloured pencils, rulers marked in centimetres, plain paper, and centimetre squared paper are used so often that they are not mentioned in the following list.

Prelude Set of pinboards—at least up to 5 × 5 lattice, rubber bands for pinboards. Spotty paper.

Chapter 3 Clock face. Cardboard and pins. Set of protractors, blackboard protractor. Cotton and drinking straws.

Chapter 4 Set of spike abaci and E.S.A. adjustable spike abacus for demonstration.

Chapter 5 Ink (for making ink devils). Sets of plane mirrors, scissors and compasses. Sheets of thin plain paper for folding, tracing paper.

Chapter 6 A ruler marked in eighths of an inch.

Chapter 7 Chalk or rope and blackboard protractor. Set of scissors, gummed paper. Set of compasses.

Chapter 10 Thin card, sellotape, glue, set of scissors. Isometric graph paper. Set of compasses. Drinking straws and pipe cleaners or cotton.

T

Prelude

Where possible, problems to which mathematics can be applied should be derived from practical situations. It would be unfortunate if all the mathematical ideas and the reasoning associated with them were to be introduced to pupils in the context of 'text book problems'. Sometimes these practical situations will be found outside the school but frequently, as in this prelude, they will consist of experiments that can be carried out within the classroom.

Our present purpose is to stimulate interest in pattern; pattern both in visual effect and in the thinking that is necessary to obtain some of the answers.

It is hoped that there will be time for pupils to experiment for themselves and to form their own conclusions. Where little progress seems to be being made, it is sometimes helpful to suggest a systematic approach; for example, Experiments 3, 4 and 5 lead to the answer to the question, 'How many triangles are there altogether?'. Notice that many of the ideas that will be discussed during the book are first mentioned in this prelude. For example, there is a brief introduction to symmetry, rotation, fractions, coordinates and polygons of various sorts.

The making of pinboards

To do the job properly either 9 mm or 12 mm ply is needed together with brass escutcheon pins. However, for the 3×3 and 5×5 it is often possible to obtain small off-cuts of wood at little or no cost. A 4 cm gap between the pins is suggested. It is sometimes possible for a set of boards to be made by some senior pupils at the end of the year after their examinations, in preparation for the coming year's intake.

Value of the pinboard

Apart from helping to form preliminary thoughts as suggested in this prelude, the boards can be used to advantage in much of the later work. There will be opportunities to use them in the chapters on coordinates and fractions as well as those on polygons, number patterns and area.

Note on the production of spotty paper

Spotty paper is just a square lattice of dots produced by an ink stencil cut on an ordinary typewriter. With experiment it will be found that a certain number of spaces with the spacer bar is equal in distance to a number of clicks of the roller ratchet.

This is obviously useful for recording pinboard patterns where often the patterns would be spoilt if the lines of ordinary squared paper were included. Spotty paper can also be used for work on coordinates and fractions.

1. THE THREE BY THREE PINBOARD

Experiment 1

Six squares are possible.

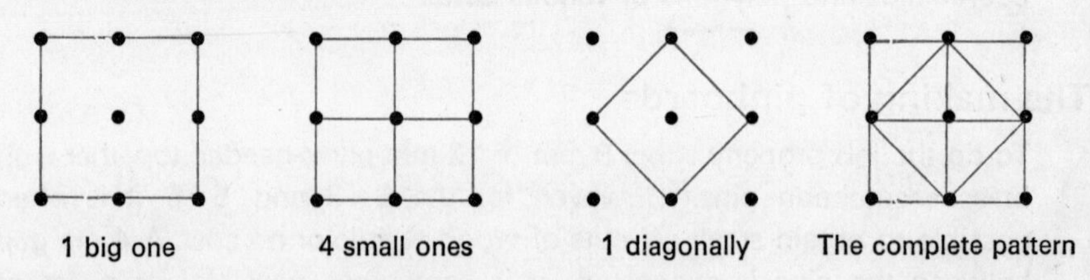

1 big one   4 small ones   1 diagonally   The complete pattern

Experiment 2

Four rectangles which are not squares are possible.

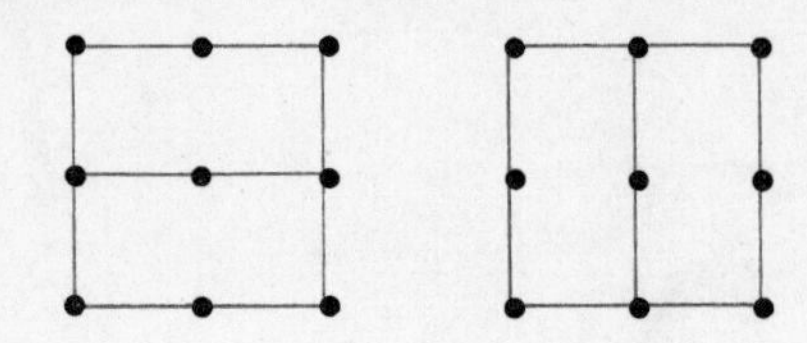

Prelude

1. THE THREE BY THREE PINBOARD

Let us see how much mathematics we can find in a simple situation. We shall need a piece of wood with nine small nails hammered into it to form the above square pattern. Use elastic bands to carry out the following experiments:

Experiment 1

How many squares can you make?
Use spotty paper or squared paper to record each separate discovery, and then show all your results on one diagram.

Experiment 2

How many rectangles can you make? Record your results.

You probably found these two experiments fairly easy. It is possible to make only six squares and four other rectangles. When we think about other shapes, such as triangles, the task of finding all of them is more difficult. We shall split this task into parts.

Experiment 3

Find triangles which have two sides equal in length like these:

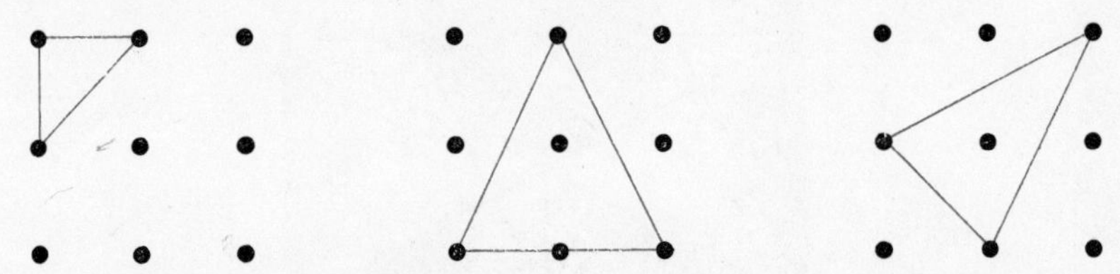

These are called *isosceles* triangles.
Record your results carefully. There are many triangles to be found.

Experiment 4

Find triangles which do not have any equal sides but which do have a square corner, such as:

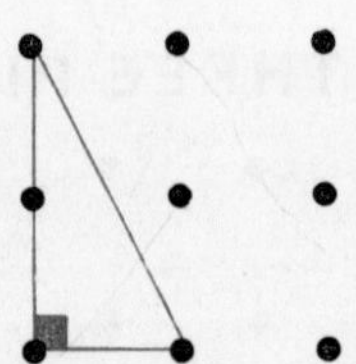

Experiment 5

Are there any other triangles?
How many triangles can you find altogether?
Can you find a triangle with all three sides the same length?

Experiment 6

These shapes are called *parallelograms*. How many can you find that have no square corners?

Experiment 7

Shapes which have four straight sides are called *quadrilaterals*. Apart from squares, rectangles and parallelograms without square corners, what other quadrilaterals can you make?

Experiment 3

Thirty-six isosceles triangles are possible.

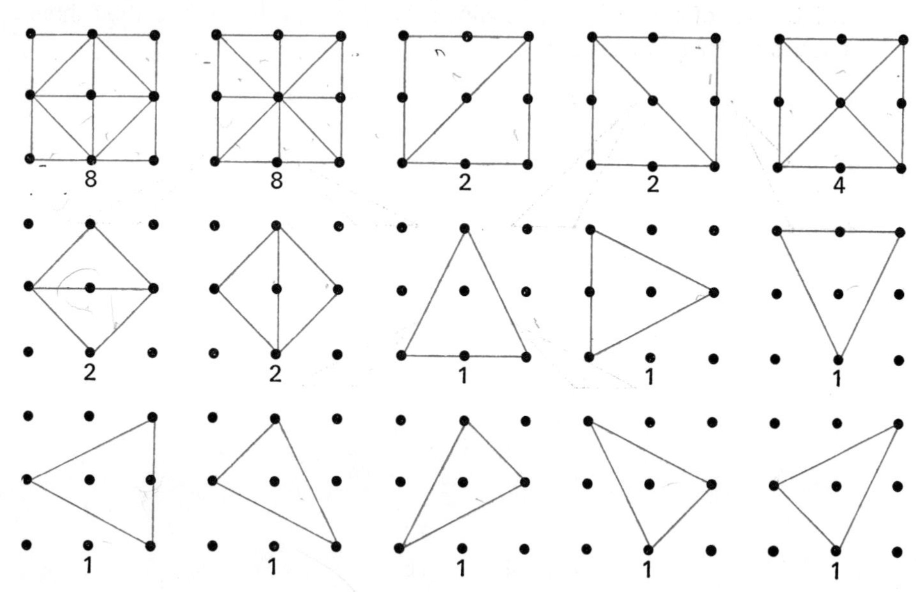

Experiment 4

Sixteen right-angled triangles (non-isosceles) are possible.

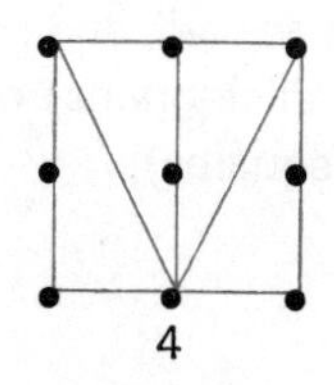

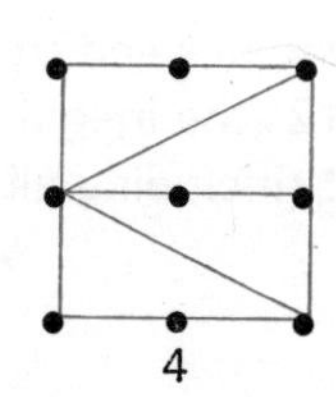

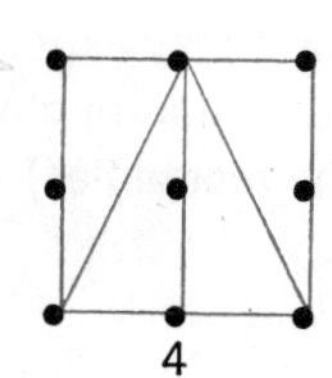

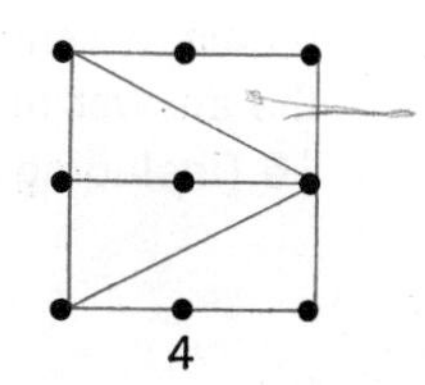

Experiment 5

There are 24 other triangles. Equilateral triangles are impossible, but there will be four of each of the following: the triangles shown and then three images under quarter-turns.

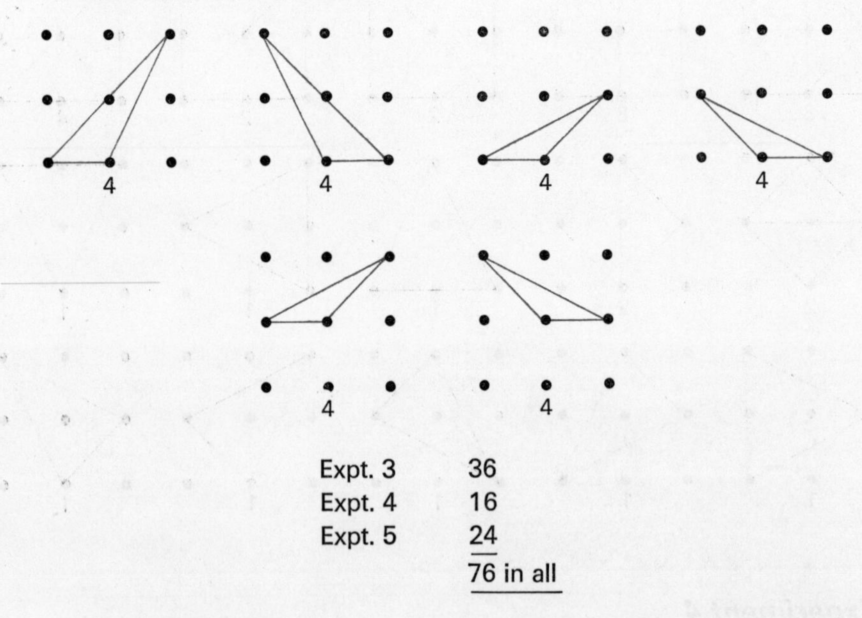

Expt. 3	36
Expt. 4	16
Expt. 5	24
	76 in all

The next questions attempt to indicate the hierarchy of quadrilaterals but this is not a matter to be stressed. If we ask for all the possible parallelograms, we should probably expect rectangles and squares as well, and so the answer to the question might be 12 (the irregular parallelograms) or 16 (including all the rectangles) or 22 (including all the squares).

Experiment 6

There are 12 parallelograms. The position of those not shown in the figure can be obtained by rotations about the centre point.

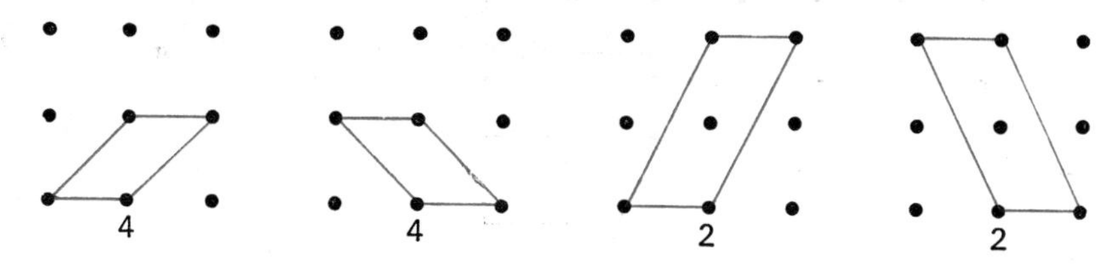

Experiment 7

Other quadrilaterals: there are 28 trapezia.

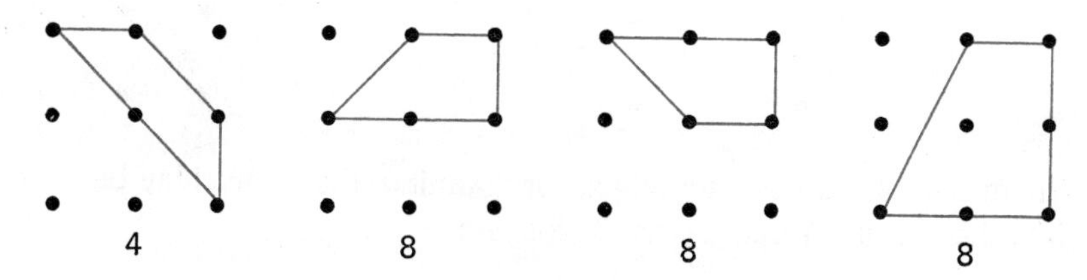

There are 4 kites and 8 arrowheads.

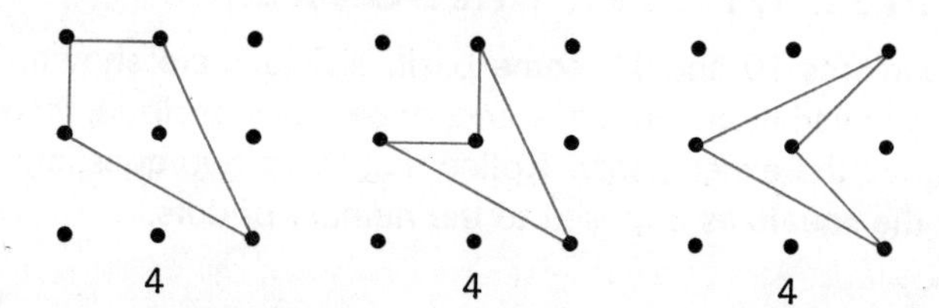

There are 16 irregular convex quadrilaterals and 16 which are not convex.

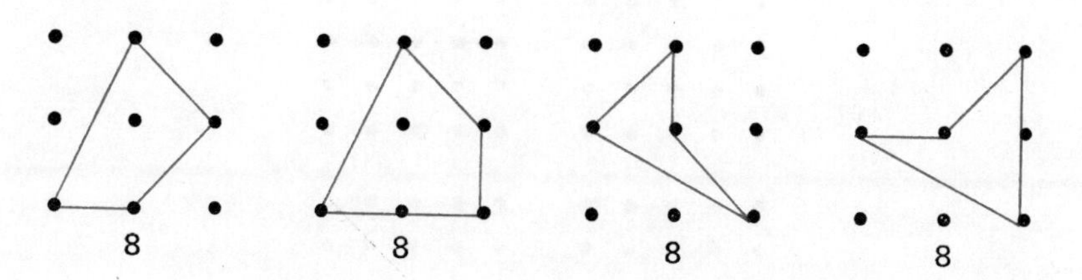

It is therefore possible to make 94 quadrilaterals on a 3×3 pinboard; 70 are convex and 24 are non-convex.

Experiment 8

Many shapes are possible with 5, 6 or 7 sides, but it is impossible to make a shape with 8 sides (unless crossed-over polygons are allowed).

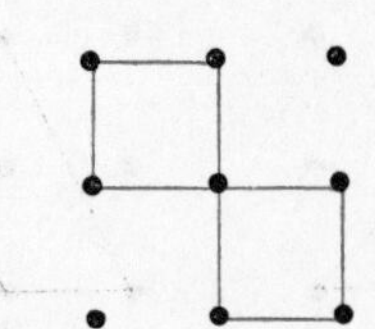

Experiment 9

Shortest route. One longest route.

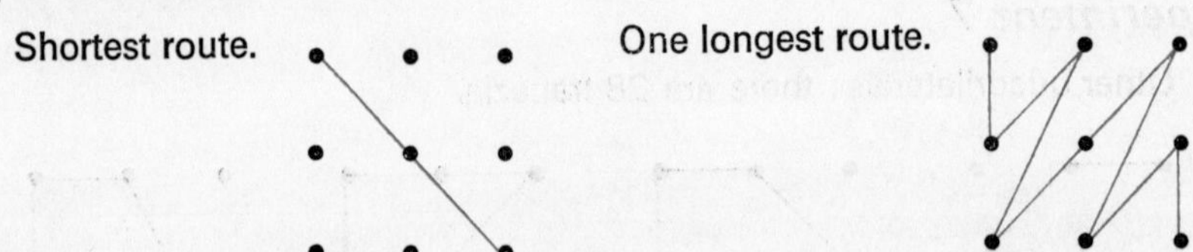

There are 389 routes altogether. For counting, the routes may be divided into those using 1 pin, 2 pins, 3 pins, etc.

2. THE FIVE BY FIVE PINBOARD

In Experiments 10 and 11, some basic divisions are shown. There are a great many variations on these and, once this is realized, there is no need to continue the experiments. Notice that these two questions refer to the area of the square as opposed to the number of dots.

Experiment 10

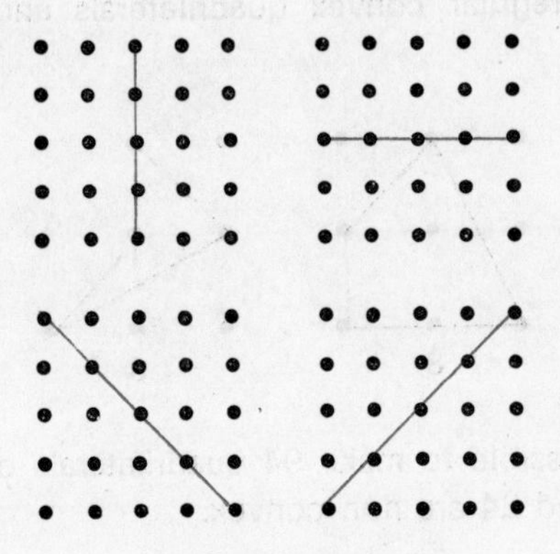

Experiment 8

Can you make shapes which have five, six or seven sides? Is it possible to make a shape with eight straight sides?

Experiment 9

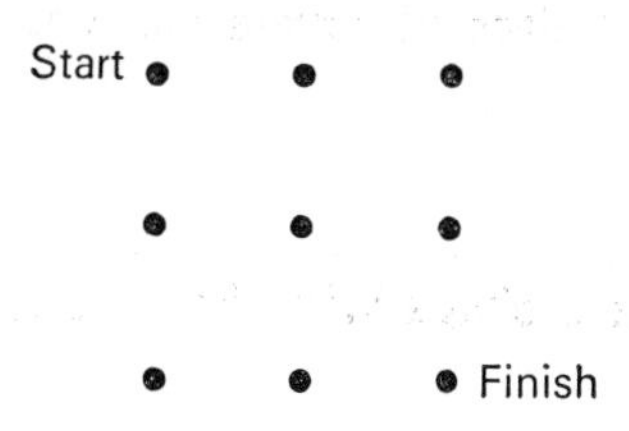

What is the shortest route from the top left corner to the bottom right? What is the longest route you can find? How many routes do you think there are? (You may use each nail only once and the bands should not cross over.)

2. THE FIVE BY FIVE PINBOARD

If, instead of nine nails, twenty-five are used, a five by five pinboard can be made. Finding all the squares, all the triangles, and so on, is now more difficult than it was with a three by three board. However, there are some easier problems which we can consider.

Experiment 10

Using one elastic band as a boundary, the board can be halved in this way, or even this.

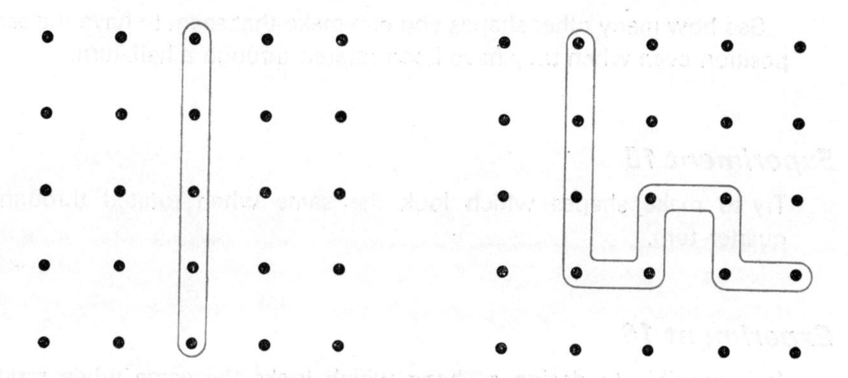

In how many different ways can you halve the board?

Experiment 11

In how many different ways can you quarter the board?

Experiment 12

How many different sizes of square can you make?

Experiment 13

See who can make a shape with the greatest number of sides.

Experiment 14

Make this shape on your pinboard. If somebody came along while you were not looking and turned the board through half of a complete turn, would you know what had happened?

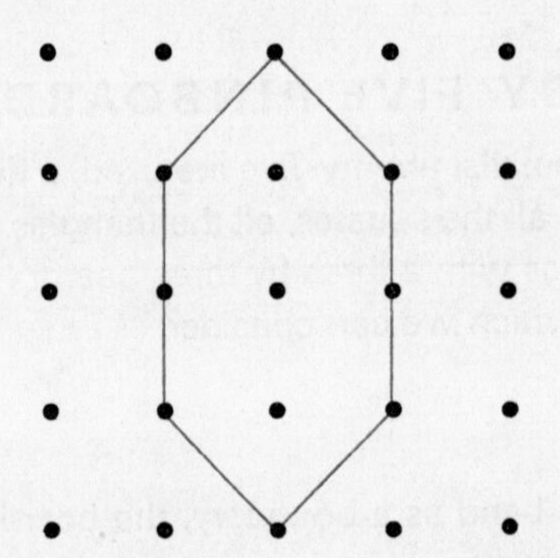

See how many other shapes you can make that seem to have the same position even when they have been rotated through a half-turn.

Experiment 15

Try to make shapes which look the same when rotated through a quarter-turn.

Experiment 16

Is it possible to design a shape which looks the same when rotated through an eighth of a turn?

Experiment 11

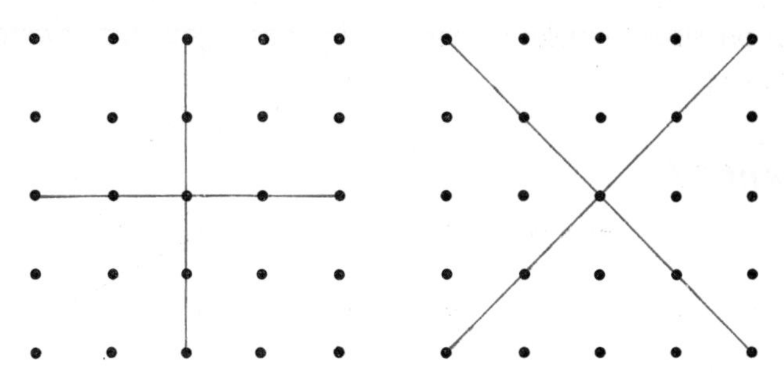

Experiment 12

There are squares of eight different sizes. (There are 16 1-unit-sided squares, 9 2-unit-sided squares, 4 3-unit-sided squares, 1 4-unit-sided square, 9 squares of side $\sqrt{2}$ units, 8 squares of side $\sqrt{5}$ units, 2 squares of side $\sqrt{10}$ units, 1 square of side $\sqrt{8}$ units; 50 altogether.)

Experiment 13

Here is a shape with 24 straight sides.

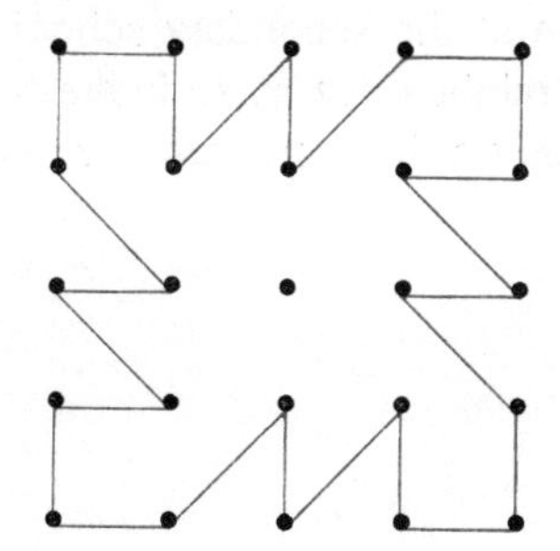

Experiment 14

A half-turn would not be noticed. A great many such shapes exist, especially if we allow cross-over patterns.

Experiment 15

Many shapes exist.

Prelude

Experiment 16

It is not possible to make a shape which has eighth-turn symmetry on a five by five pinboard.

Experiment 17

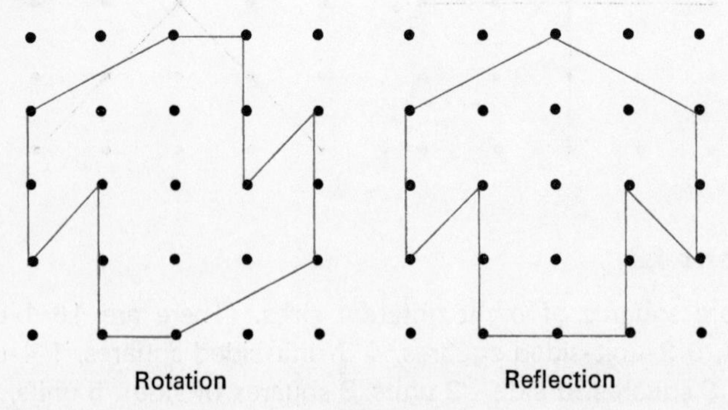

Experiment 18

Many shapes exist.

Experiment 19

Pupils frequently arrive at the secondary school having met coordinates and this short section might serve as an interesting reminder of the basic idea before they embark on the more formal work to follow.

Experiment 17

Somebody was in a hurry and left this shape on his board saying that the other half was the same anyway.

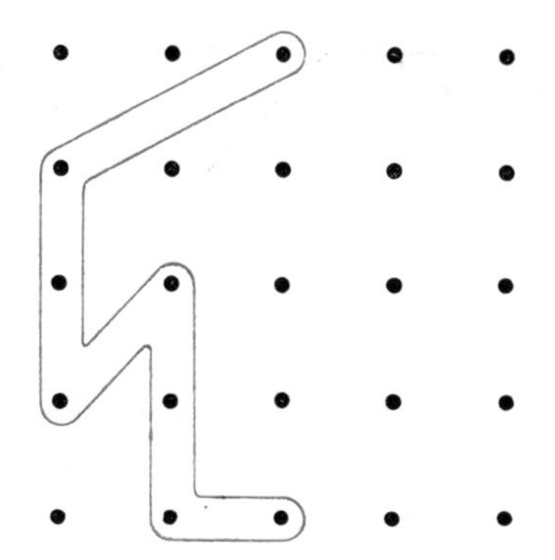

Can you complete the shape? Is there more than one possible answer?

Experiment 18

Design other shapes which could have been given for Experiment 17. Work with a partner. Make your half and ask your partner to complete the pattern.

Experiment 19

How could you instruct someone over the telephone if you wanted him to reproduce this shape?

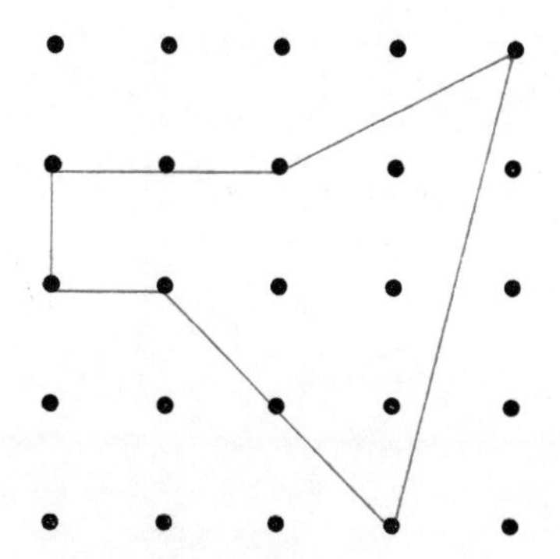

Give each corner of your figure a letter and then mark the bottom left-hand corner nail as the starting point, S.P.

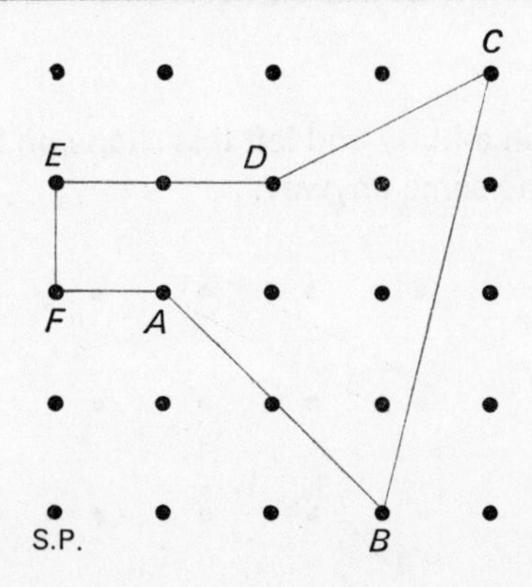

If the person on the other end of the telephone has a board numbered in the same way, you could then give instructions like this:

'Point A; 1 along and 2 up.' 'Point B; 3 along and 0 up.'

'Point C; 4 along and 4 up.' 'Point D; 2 along and 3 up.'

'Point E; 0 along and 3 up.' 'Point F; 0 along and 2 up.'

If you also agree between yourselves that you will always give the along-number first and the up-number second, the instructions become much simpler and might now be: A is (1, 2); B is (3, 0); C is (4, 4); D is (2, 3); E is (0, 3); F is (0, 2). Work in pairs. Take it in turns to make a shape without letting the other person see it and give instructions using this method. Compare the shapes you have obtained.

The conventional method of indicating coordinates, by number along the axes, will be developed in Chapter 2. The informal method shown here gives the conventional results.

T

1. Number Patterns

In this chapter, visual methods are used to stimulate an investigation into various sets of counting numbers. Primes and composite numbers (rectangle numbers) are considered in this chapter and in Chapter 8, and also more regular sequences such as odd, square, triangle and pyramid numbers.

Where possible, pupils should be encouraged to register the visual patterns, to describe these patterns in numbers and then to search for connections between the terms of the sequences using, for example, first-order differences.

There is to be no chapter on sets in this book. The ideas and notation of sets will be introduced within various chapters as they are required. If a number belongs to a set, we say that it is a 'member' or an 'element' of that set.

Here we introduce the methods of defining sets by enumeration and description using curly brackets. (In a later book, we shall introduce the idea of sets in a non-numerical context, but, to begin with, we shall consider only sets of counting numbers and, in the Coordinates Chapter, points.)

1. PATTERNS OF DOTS

In this section, attention is drawn to the possibility that dots can be arranged in a manner which has numerical significance.

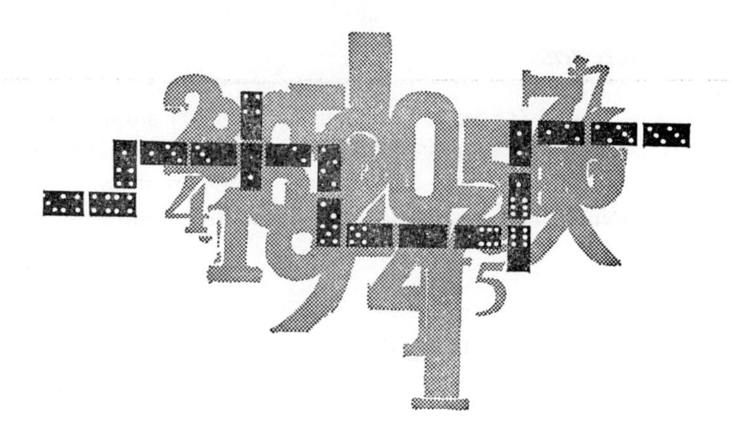

1. Number patterns

1. PATTERNS OF DOTS

How many dots are there in each pattern?

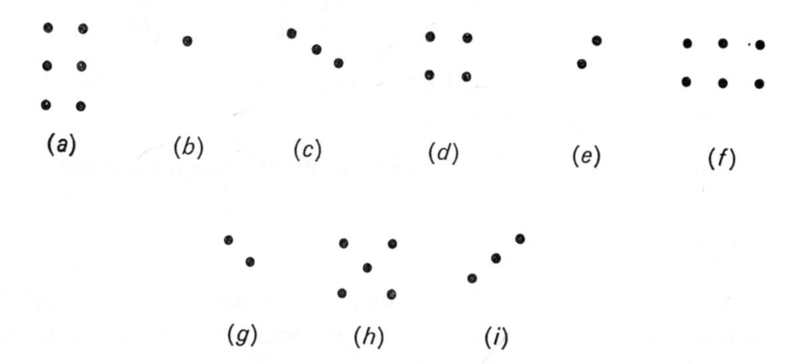

Did you have to count them or did you know the answers?

Make some dot patterns like these and show them quickly to your neighbour. Can he tell how many there are? Is it easier for him if the dots form a regular pattern?

Here are some of the ways we can arrange six dots to form a pattern.

(a) (b) (c) (d) (e)

Patterns (*a*) and (*b*) are rectangular, (*c*) and (*d*) are like straight lines and (*e*) is triangular. Which do you find the easiest to recognize as six?

We shall consider (*a*) and (*b*) to be the same pattern; also (*c*) and (*d*). The arrangement is the same in each case: its position on the page is different.

2. RECTANGLE NUMBERS

Any number that can be shown as a rectangular pattern of dots is called a *rectangle number*. Fifteen is a rectangle number for it can be shown as

• • • • •
• • • • •
• • • • •

Exercise A

1 Try these numbers and find out whether they will make rectangular patterns:

(*a*) 8; (*b*) 3; (*c*) 10; (*d*) 18;
(*e*) 7; (*f*) 12; (*g*) 21; (*h*) 20.

2 Now try (*d*), (*f*) and (*h*) again and see whether you can make a different rectangular pattern.

3 Can you find a number that has three or more rectangular patterns?

2.1 Finding the patterns

You will have found that you can only make rectangular patterns for certain numbers. All these rectangle numbers can be divided by numbers other than themselves and one.

We can see, for example, that twelve can be set out as

• • • •
• • • • or • • • • • •
• • • • • • • • • •

These patterns show 12 as 3×4 or as 2×6.

21 can be set out as

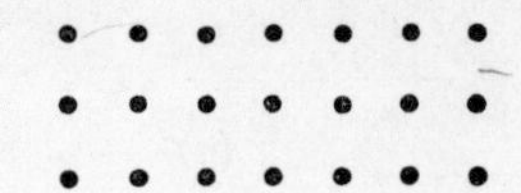

We know that $21 = 3 \times 7$.

2. RECTANGLE NUMBERS

Any number which is a composite number, i.e. which has factors, can be shown as a rectangle of dots or perhaps as several different rectangles. To distinguish between composite and prime numbers, we shall not consider a single row or column of dots to be a rectangle.

Exercise A

1 (*a*) Yes; (*b*) no; (*c*) yes; (*d*) yes;
(*e*) no; (*f*) yes; (*g*) yes; (*h*) yes.

2 (*d*) could be 2×9 or 3×6;
(*f*) could be 2×6 or 3×4;
(*h*) could be 2×10 or 4×5.

3 Examples are 24, 30, 36, etc. multiples of 6

2.1 Finding the patterns

In this chapter we are not interested in distinguishing between three rows of four or four rows of three when representing twelve; we shall consider them to be the same pattern. Nevertheless, if the question is raised, this does give an opportunity for pointing out that multiplication of counting numbers is commutative (i.e. $ab = ba$ where a and b are counting numbers).

Sets are said to be 'equal' if they contain the same elements. Thus {even prime numbers} = {the first even number} because the element in both these sets is the same.

The systematic approach to finding a set of factors of a number should be encouraged as it saves unnecessary labour. At this stage, there is no harm in letting pupils consult copies of their multiplication tables to help with this work; if necessary, extra examples (using numbers less than 100) could be set.

Of course $12 = 3\times 4 = 4\times 3$. Though these patterns are the same, we shall make a point of stating the number of rows first, so

• • • •
• • • •
• • • •

will be called a 3×4 rectangular pattern,

• • •
• • •
• • •
• • •

a 4×3 rectangular pattern.

2, 3, 4 and 6 are called *factors* of 12.
3 and 7 are factors of 21.
Every rectangle number has its own set of factors.

Example 1

Find the factors of 48.

$$48 = 2\times 24 \qquad 48 = 6\times 8 \qquad 48 = 3\times 16$$
$$48 = 4\times 12 \qquad 48 = 1\times 48$$

and even though the last two numbers do not give us a rectangular pattern, they do multiply together to give 48, so we include them in the set of factors.

The factors of 48 are 1, 2, 3, 4, 6, 8, 12, 16, 24 and 48.

When we are sure that we have found the whole set of factors of a number, we can write them like this:

$$\{\text{factors of } 48\} = \{1, 2, 3, 4, 6, 8, 12, 16, 24, 48\}.$$

The curly brackets are a short way of writing, 'The set of'.

{factors of 48} is read, 'The set of factors of 48', and this *describes* or *names* the set we are referring to.

{1, 2, 3, 4, 6, 8, 12, 16, 24, 48} is read 'The set 1, 2, 3, 4, 6, 8, 12, 16, 24, 48,' and this *lists* the *members* or *elements* of the set.

Example 2

Find {factors of 30}.

$$30 = 1\times 30 \qquad 30 = 2\times 15$$
$$30 = 3\times 10$$
$$30 = 5\times 6$$

$$\{\text{factors of } 30\} = \{1, 2, 3, 5, 6, 10, 15, 30\}.$$

In Example 1, we found quite a large number of factors for 48 and Example 2 shows us that 30 also has quite a lot. This could mean that with some numbers we may find it difficult to know when we have actually found all of them. The method of Example 2 will help us to do this.

We know that 1 will be a factor of all numbers, so we write that down first. Then we start with the next counting number, 2, and find out whether it is a factor. From there, go on to three, four, etc., until you have found them all.

Exercise B

1 Find the set of factors of each of the following numbers:

(*a*) 15; (*b*) 27; (*c*) 32; (*d*) 36; (*e*) 56;
(*f*) 63; (*g*) 72; (*h*) 81; (*i*) 24; (*j*) 33.

How did you know when to stop trying new numbers? Look at 36 again. When we try numbers to see whether they are factors we get: 1×36; 2×18; 3×12; 4×9; 6×6, and we have now found them all. If we did try to go on, the next result would be 9×4 and we already have these two factors. This shows that we can stop as soon as we come to two factors which are the same, or which repeat the previous pair. Look at 30 again. We get: 1×30; 2×15; 3×10; 5×6, and the next one will be 6×5, two numbers we have already.

3. SQUARE NUMBERS

3.1 A new pattern

We have already seen that some numbers have more than one pattern. Now we shall look at some special ones.

Example 3

Show 16 as a rectangle of dots in as many ways as possible.

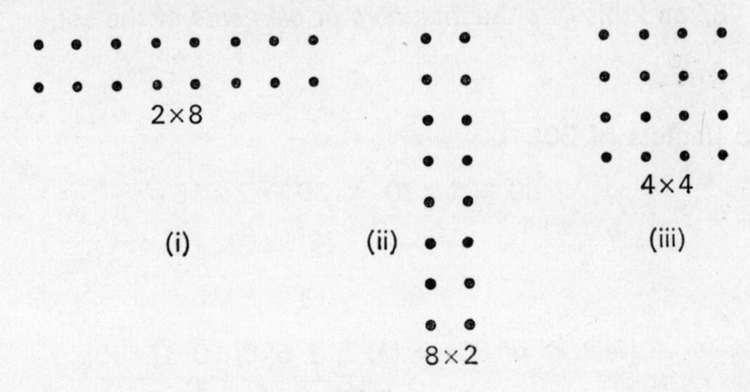

Two of the answers look very like those you have met already, but the third is a little different. There are the same number of rows as columns and so the pattern is a square of dots. Try to find three other numbers which can be represented by a square of dots. Such numbers are called *square numbers.*

To do this we had to find numbers which had two equal factors: $7 \times 7 = 49$; $6 \times 6 = 36$; $9 \times 9 = 81$; and a special one, $1 \times 1 = 1$.

'7×7' is often written as '7^2' which we read as '7 squared'. In the same way $16 = 4 \times 4 = 4^2$.

If we start at one, these numbers make a very interesting series of patterns:

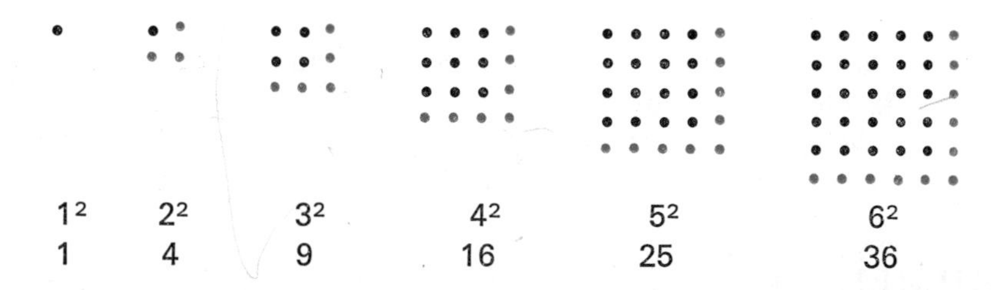

What are the next two numbers of this kind?

Notice that we have one square number which is not a rectangle number. Which is it?

We have shown the first six square numbers as a pattern of red and black dots. The red dots have been used to show how one square number is built up from the one before it.

Exercise C

1 (*a*) Draw a dot pattern for 3^2. By adding more dots, make it into a pattern for 4^2. How many dots did you add?

(*b*) Using the same method, make 5^2 into 6^2. How many extra dots did you use?

(*c*) Make 8^2 into 9^2. How many extra dots are needed?

2 Copy and complete the following:

$$2^2 = 1^2 + ?,$$
$$3^2 = 2^2 + 5,$$
$$4^2 = 3^2 + ?,$$
$$5^2 = ? + ?.$$

3 What relation is there between a square number and the square number before it?

4 The figure shows the square number 25 with lines drawn to divide the dots into different sets.

(*a*) By looking at the numbers of dots between the lines, and the number of dots in the square cut off by a particular line, complete the following:

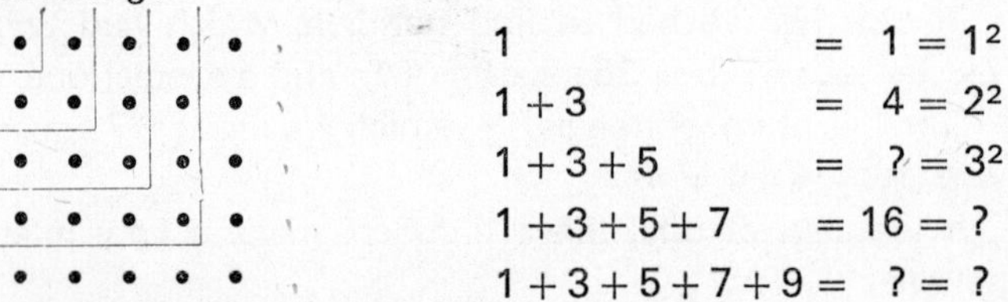

$$1 = 1 = 1^2$$
$$1+3 = 4 = 2^2$$
$$1+3+5 = ? = 3^2$$
$$1+3+5+7 = 16 = ?$$
$$1+3+5+7+9 = ? = ?$$

(*b*) Now write down the next line.

5 Use the pattern of numbers in Question 4 to work out the sum of the first ten odd numbers.

What is the connection between the *number* of odd numbers summed and the *number* which is squared to equal their sum?

SUMMARY

A *factor* is a counting number which divides a whole number of times into another counting number.

Each *rectangle number* has more than two *factors*.

A *square number* is the product of two *equal factors*.

Miscellaneous Exercise D

1 (*a*) Draw as many rectangular patterns as you can for each of the following numbers:

(i) 18; (ii) 24; (iii) 30; (iv) 40; (v) 60.

(*b*) Now write down the set of factors of each of these numbers in turn.

2 (*a*) Draw as many rectangular patterns as you can for:

(i) 25; (ii) 36; (iii) 49; (iv) 64.

(*b*) Write down the set of factors of each of these numbers in turn.

3 (*a*) Use the relation between the numbers of odd numbers and the square numbers to find:

(i) $1+3+5+7 = ?$,

(ii) $1+3+5+7+9+11 = ?$,

(iii) $1+3+5+7+9+11+13+15+17+19 = ?$,

(iv) $1+3+5+7+9+\ldots+37+39 = ?$.

(*b*) What is $21+23+25+27+\ldots+37+39$?

4 (*a*) $9=3^2$; $16=4^2$; $25=5^2$;
(*b*) $1+3+5+7+9+11=36=6^2$.

5 $1+3+5+7+9+11+13+15+17+19=100$;
i.e. the sum of the first 10 odd numbers $=10^2=100$.

Miscellaneous Exercise D

1 (*a*) Considering 3×4 to be the same as 4×3, we have:
(i) $2\times9, 3\times6$;
(ii) $2\times12, 4\times6, 3\times8$;
(iii) $2\times15, 3\times10, 5\times6$;
(iv) $2\times20, 4\times10, 5\times8$;
(v) $2\times30, 3\times20, 4\times15, 5\times12, 6\times10$.

(*b*) (i) {factors of 18} = {1, 2, 3, 6, 9, 18};
(ii) {factors of 24} = {1, 2, 3, 4, 6, 8, 12, 24};
(iii) {factors of 30} = {1, 2, 3, 5, 6, 10, 15, 30};
(iv) {factors of 40} = {1, 2, 4, 5, 8, 10, 20, 40};
(v) {factors of 60} = {1, 2, 3, 4, 5, 6, 10, 12, 15, 20, 30, 60}.

2 (*a*) (i) 5×5;
(ii) $2\times18, 3\times12, 4\times9, 6\times6$;
(iii) 7×7;
(iv) $2\times32, 4\times16, 8\times8$.

(*b*) (i) {factors of 25} = {1, 5, 25};
(ii) {factors of 36} = {1, 2, 3, 4, 6, 9, 12, 18, 36};
(iii) {factors of 49} = {1, 7, 49};
(iv) {factors of 64} = {1, 2, 4, 8, 16, 32, 64}.

3 (*a*) (i) 16; (ii) 36; (iii) 100; (iv) 400.
(*b*) 300.

4 Top row: 100th. Bottom row: 99.

Each term in the bottom row is twice the corresponding number in the top row less one. The 5000th odd number is 9,999.

5 (*a*) 5; 7; 9; 11.

(*b*) The difference between:

5^2 and	4^2 is the	5th	odd number	which	is	9;
8^2 "	7^2 " "	8th	"	"	" "	15;
11^2 "	10^2 " "	11th	"	"	" "	21;
11^2 "	10^2 " "	11th	"	"	" "	21;
100^2 "	99^2 " "	100th	"	"	" "	199.

(*c*) (i) $24^2 - 23^2 = 2 \times 24 - 1 = 47$;

(ii) $24^2 = 23^2 + 47 = 529 + 47 = 576$.

6 (*a*)

$$1+2+3+4+3+2+1 = 4^2$$
$$1+2+3+4+5+4+3+2+1 = 5^2$$

(*b*)

$$1+2+3+4+5+6+5+4+3+2+1 = 6^2$$
$$1+2+3+4+5+6+7+6+5+4+3+2+1 = 7^2$$
$$1+2+3+4+5+6+7+8+7+6+5+4+3+2+1 = 8^2$$

4 Here are the first, second, .. odd numbers in order:

1st	2nd	3rd	4th	5th	...	50th	...	?
1	3	5	7	9	...	?	...	199

What numbers do the question marks stand for?
What is the 5000th odd number?

5 (*a*) What are the differences 3^2-2^2; 4^2-3^2; 5^2-4^2; 6^2-5^2?

(*b*) Complete the following:

The difference between

3^2	and	2^2	is	the	3rd	odd	number	which	is 5;
5^2	"	4^2	"	"	?	"	"	"	" ?;
8^2	"	?	"	"	8th	"	"	"	" ?;
11^2	"	?	"	"	?	"	"	"	" ?;
?	"	10^2	"	"	?	"	"	"	" ?;
?	"	?	"	"	?	"	"	"	" 199.

(*c*) $23^2=529$. Use the ideas of (*b*) to write down:

(i) the difference between 24^2 and 23^2;

(ii) 24^2.

6 The square of dots 2^2 can be split up by sloping lines like this

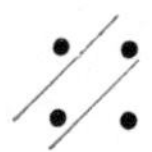

so you can see that $2^2=1+2+1$. In the same way

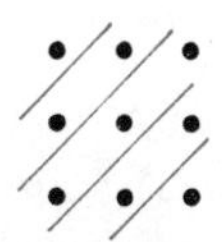

gives $3^2=1+2+3+2+1$. Use this pattern to complete the following:

(*a*)

$$1 = 1^2$$
$$1+2+1 = 2^2$$
$$1+2+3+2+1 = 3^2$$
$$1+2+3+4+3+2+1 = ?$$
$$1+2+3+?+?+?+?+?+? = ?$$

(*b*) Write down the next three lines of the pattern.

2. Coordinates

1. BATTLESHIPS

You have probably played the game of Battleships before. First of all you want your fleet.

1 battleship (B)

2 cruisers (C)

3 destroyers (D)

4 motor torpedo boats (M)

Next you will need to make a plan of your fleet. To do this, draw on squared paper two blocks of 81 squares as shown in Figure 1.

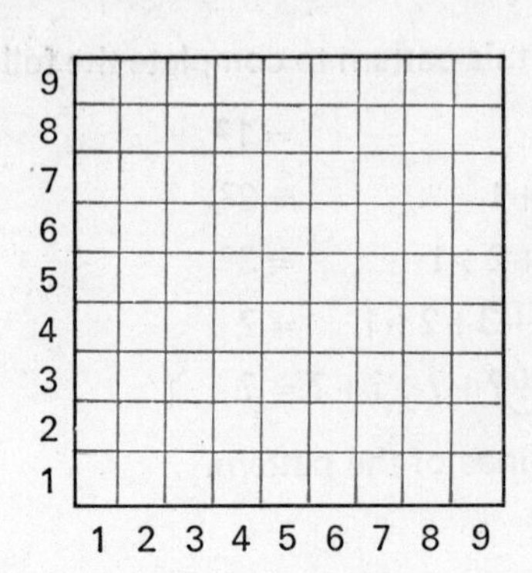

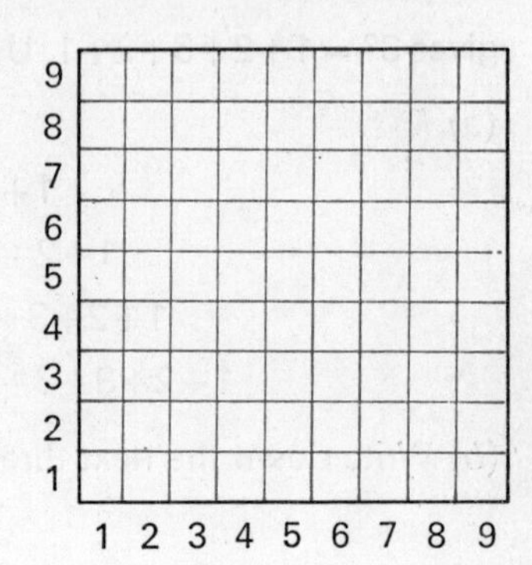

Fig. 1

T

2. Coordinates

1. BATTLESHIPS

In this section, we introduce the idea that an ordered pair of numbers can be used to label a position on a plane. Here, the ordered pair indicates a space and not a point; the latter development comes in the next section. Though 'battleships' is merely a game, it should not be omitted as the idea is central to all work with coordinates. Everybody should be given time to understand the method.

The teacher may find it advisable, before the class divides into pairs in order to play the game, to start a game between two pupils on the blackboard. In this way, those unfamiliar with the rules can learn them.

Batts

2. MAPS

In this section we use an ordered pair to define a point rather than a region. Both applications occur in mathematics, the former more than the latter. The context usually indicates which is being used.

The later questions in the following exercise introduce the idea that coordinates of a point need not be whole numbers.

One of these blocks will be for your own fleet, the other to mark your shots at the enemy's fleet. Mark the position of each ship of your own fleet as you wish. Remember that no ship can be bent, nor can two ships touch each other and that you must keep your plan secret from the enemy.

Here in Figure 2 is a game that has already been started. Each player has two plans as shown, one marked with his own fleet, and the other empty for marking his own shots at the enemy.

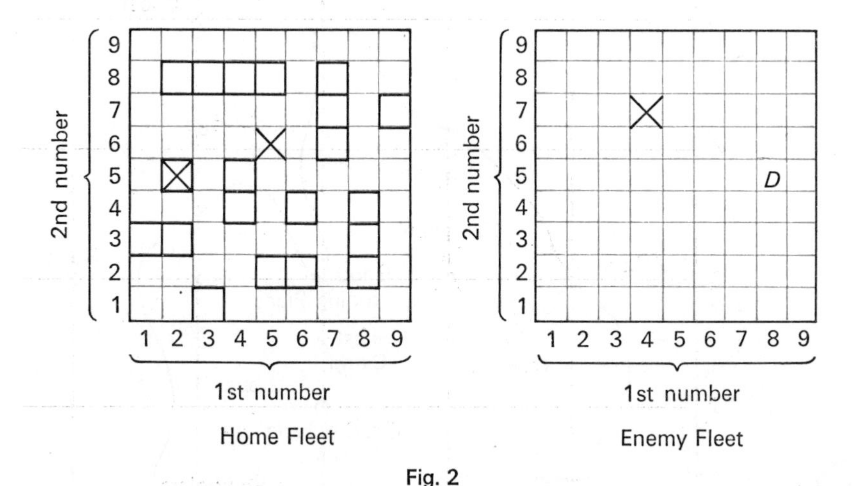

Fig. 2

Each player 'shoots' at the enemy by choosing two numbers which indicate the square on the plan in which his shot lands. The enemy shot at (2, 5) and has hit one of the Home Fleet's M.T.B.s. You must report this to the enemy so that he can mark it on his plan. The return shot at (4, 7) has hit nothing. The enemy then shot at (5, 6) and has drawn a blank. The Home Fleet's shot at (8, 5) has hit an enemy destroyer. It was marked D when the enemy admitted that his destroyer had been hit.

Now try the game with a friend in your class. Decide who will go first and then take it in turns to fire shots. The game continues until one fleet is completely destroyed.

2. MAPS

Figure 3 shows a map of 'Skull Island'. The map is divided into squares like the map you used for your fleet in the battleship game. Look at the bottom and the left-hand side of the map. There are numbers there as for battleships, but the numbers on this map have a different meaning. In the game of battleships we numbered the spaces. On the map, it is the actual lines that are numbered. On our map, the lines show the number of kilometres east and north of a fixed point—the Old Wreck.

With the map we again describe the position of something by using two numbers. Do you think that the order in which this pair is written matters?

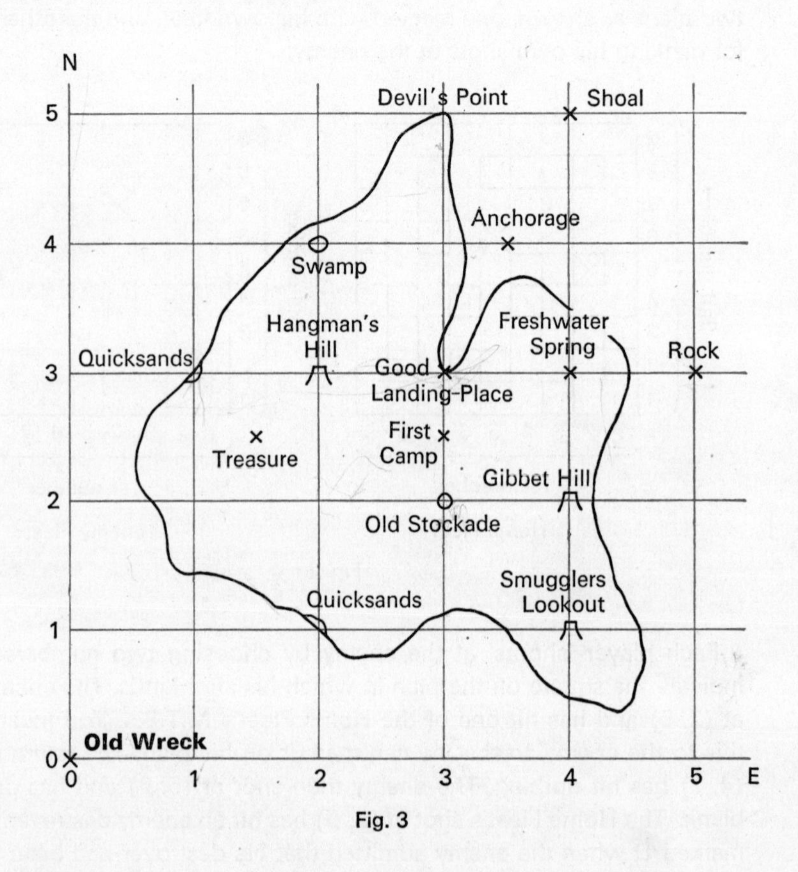

Fig. 3

It is usual to put the number of kilometres east first and the number of kilometres north second. So (4, 2) means four kilometres east and two kilometres north and therefore shows Gibbet Hill.

Exercise A

1 What places on the map are shown by the following pairs of numbers?

(*a*) (0, 0); (*b*) (3, 3); (*c*) (2, 1); (*d*) (3, 2);
(*e*) (3, 5); (*f*) (1, 3); (*g*) (5, 3); (*h*) (2, 3).

Exercise A

1 (*a*) Old Wreck; (*b*) Good Landing-Place;
(*c*) Quicksands; (*d*) Old Stockade; (*e*) Devil's Point;
(*f*) Quicksands; (*g*) Rock; (*h*) Hangman's Hill.

2 (*a*) (4, 1).
(*b*) (4, 5).
(*c*) (2, 4).
(*d*) (4, 3).
(*e*) $(3, 2\frac{1}{2})$.
(*f*) $(3\frac{1}{2}, 4)$.
(*g*) $(1\frac{1}{2}, 2\frac{1}{2})$.

3. PLOTTING POINTS

This follows directly from the previous section. Knowledge of the correct order of the coordinates is strengthened by the use of the phrases 'first name' and 'second name'.

Note that the lines $x = 0$ and $y = 0$ are called just 'axes' and not 'x-axis' and 'y-axis'. It is advisable to avoid these names as they only lead to confusion between 'the line $x = 0$' and 'the x-axis', for example.

$$C(4, 5),\quad D(1, 6),\quad E(6, 4),\quad F(3, 0).$$

The first three questions of the following exercise might well be tackled orally, using the expressions 'first name', 'second name' and 'full name' together with the word 'coordinates' as long as the pupils find it necessary.

2 What pairs of numbers give the positions of the following places?

(*a*) Smugglers' Lookout.
(*b*) Shoal.
(*c*) Swamp.
(*d*) Freshwater Spring.
(*e*) First Camp.
(*f*) Anchorage.
(*g*) Treasure.

3 Make up your own map. Choose a fixed point and draw the lines as in Figure 3. Mark in some interesting places and write down the pairs of numbers which give their positions.

3. PLOTTING POINTS

Have a look at Figure 4. This is rather like the pattern of lines on the map in Figure 3. Exactly as on the map, points can be indicated by using two numbers. In the same way as you yourself have a first and second name, so each point has a first and second name. Look at the point *A* in Figure 4. Its first name is 2 and its second name is 3. So its full name is (2, 3). Look at the point *B*. Its first name is 5 and its second name is 2. So its full name is (5, 2).

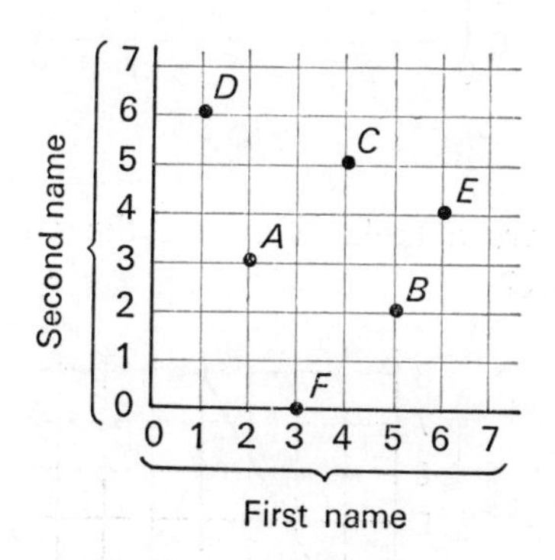

Fig. 4

What are the full names of *C, D, E, F*?

We usually call the full name of a point its '*coordinates*'. So the coordinates of *A* are (2, 3) and the coordinates of *B* are (5, 2).

What are the coordinates of *C, D, E, F*?

Notice that it is usual to take the first name from the set of numbers across the page. The second name comes from the set of numbers up the page.

Exercise B

1

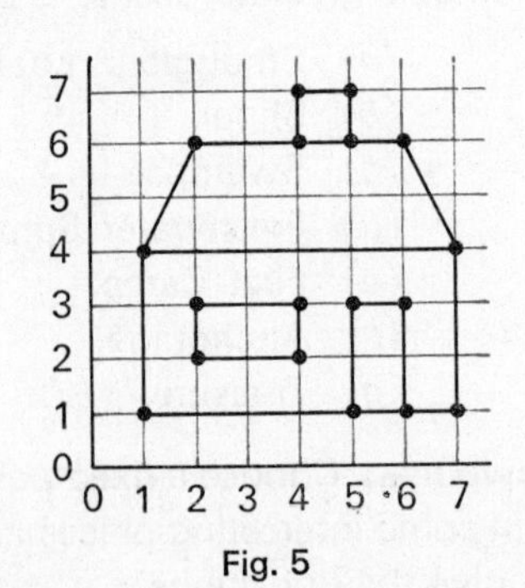

Fig. 5

Figure 5 shows a house. What are the coordinates of

(*a*) the top of the chimney; (*b*) the corners of the window;
(*c*) the corners of the door; (*d*) the corners of the roof?

2

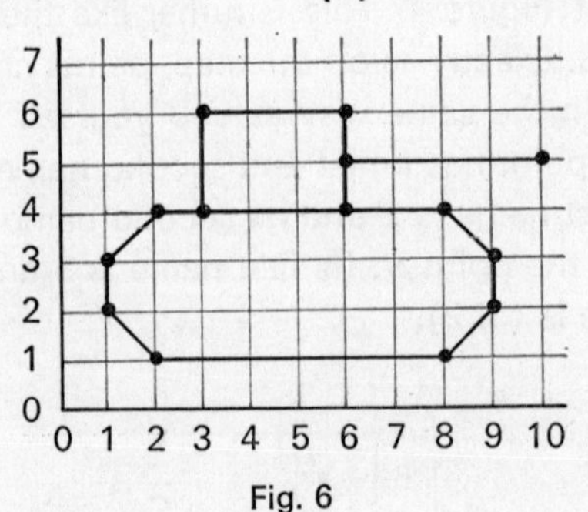

Fig. 6

Figure 6 shows a tank. What are the coordinates of the points marked in black?

3

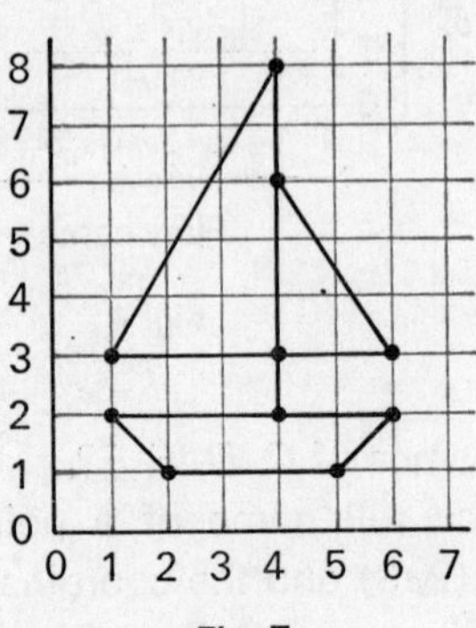

Fig. 7

Figure 7 shows a sailing boat. What are the coordinates of the points marked in black?

Exercise B

1 (*a*) (4, 7), (5, 7);
(*b*) (2, 2), (2, 3), (4, 2), (4, 3);
(*c*) (5, 1), (5, 3), (6, 1), (6, 3);
(*d*) (1, 4), (2, 6), (6, 6), (7, 4).

2 (2, 1), (8, 1), (9, 2), (9, 3), (8, 4), (6, 4), (6, 5), (10, 5), (6, 6), (3, 6), (3, 4), (2, 4), (1, 3), (1, 2).

3 (1, 2), (2, 1), (5, 1), (6, 2), (4, 2), (4, 3), (6, 3), (4, 6), (4, 8), (1, 3).

Coordinates

4

(a)

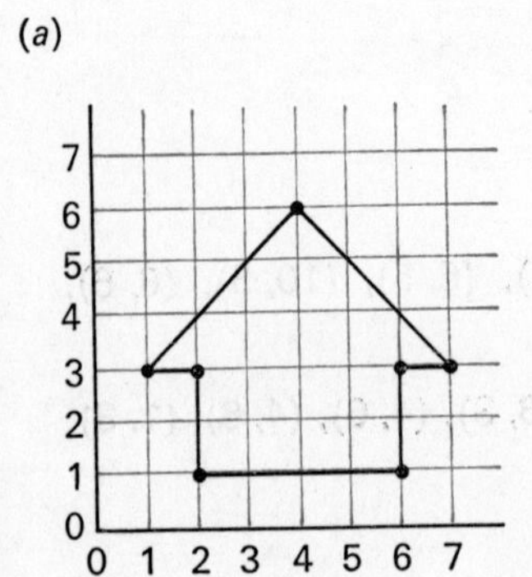

(b)

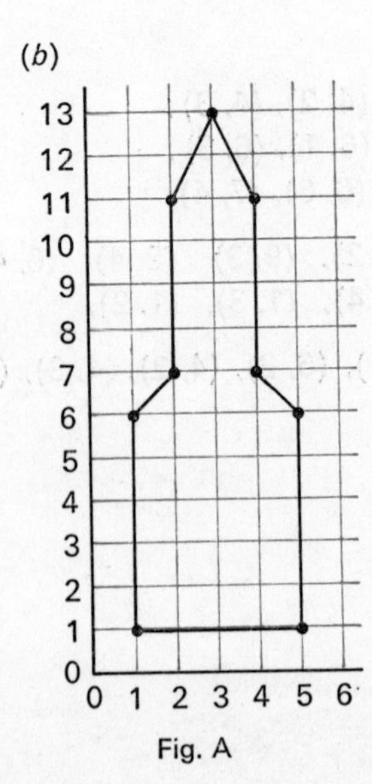

(c)

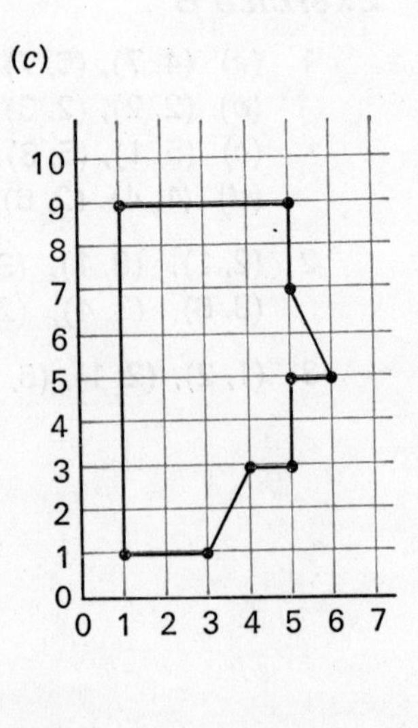

Fig. A

In (c), at (4, 7) there should be an eye.

5

(a)

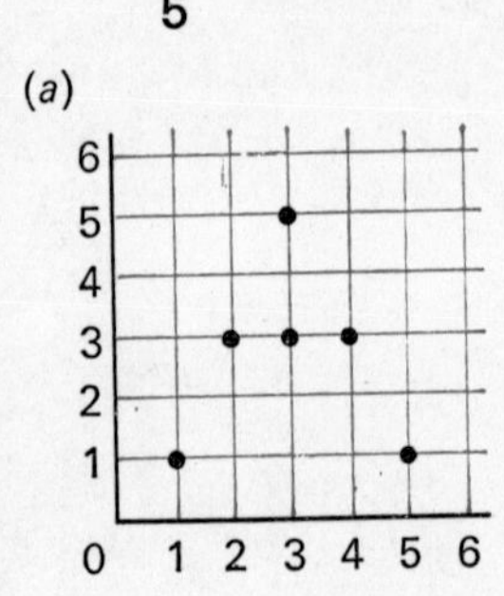

(b)

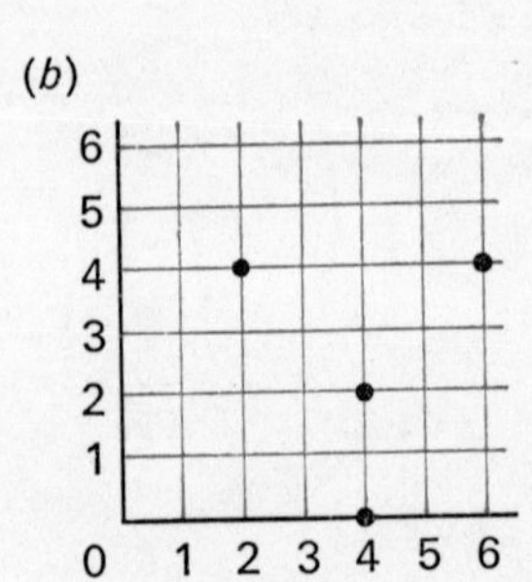

(c)

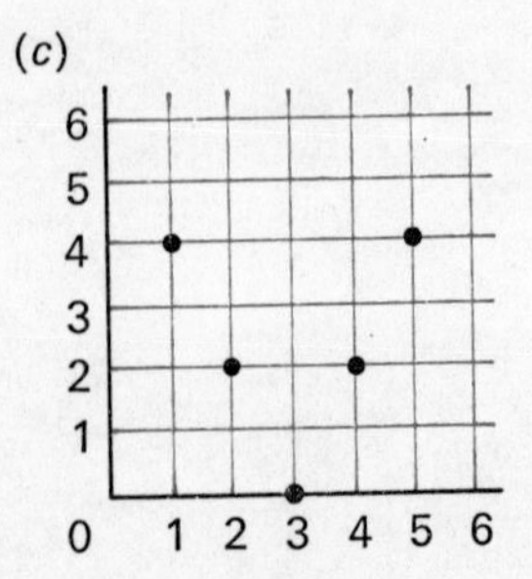

(d)

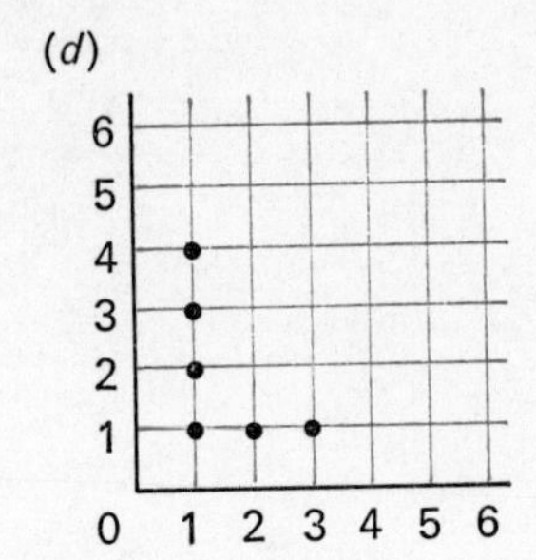

(e)

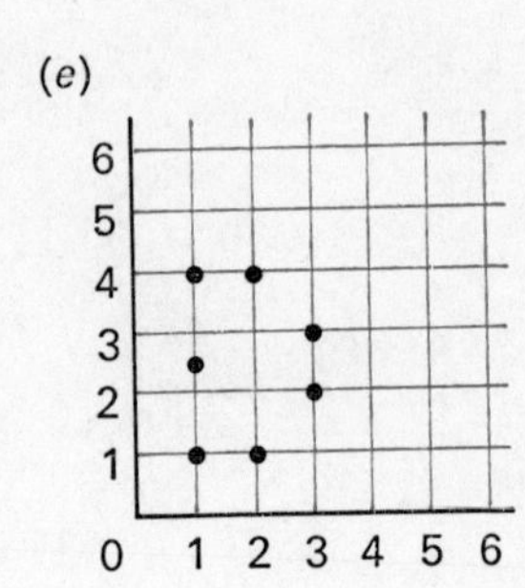

Fig. B

6 (*a*) (1, 5); (*b*) (3, 2); (*c*) (4, 4);
(*d*) (1, $2\frac{1}{2}$); (*e*) (3, 1); (*f*) ($4\frac{1}{2}$, 5);
(*g*) ($5\frac{1}{2}$, $3\frac{1}{2}$); (*h*) (2, 5); (*i*) (4, $3\frac{1}{2}$).

It is best for the pupils to express a rule for finding the coordinates of the mid-point in their own language. Afterwards different rules can be compared, and perhaps streamlined.

The mathematical formula for the mid-point of a line segment from (x_1, y_1) to (x_2, y_2) is

$$\left(\frac{x_1+x_2}{2}, \frac{y_1+y_2}{2}\right),$$

but no attempt should be made to obtain this as an algebraic statement.

7 (3, 5) the mid-point of the first line.

8 This question might well be done after Question 4 or Question 5, if the teacher feels that that is a better place for it. The preparatory drawing of pictures takes rather a long time and a suitable homework might be for each pupil to prepare several pictures and then read out the coordinates to his neighbour in the next class period.

4. LINES

The number line will be mentioned in Chapter 4. Here, it is used, but not mentioned. The idea is to build up lines, first as sets of points associated with the integers, and then to add some intermediate points associated with the rational numbers. Pupils will tend to assume that rational numbers will fill the whole line but, in fact, the real numbers—rationals together with irrationals—are required (a point that need only be mentioned if it is raised in class).

An alternative approach to the one in the book, which can also be used as an introduction, is to draw a grid on the board and ask for all points whose first name is 3 to be marked on it. At first the diagram will look like this.

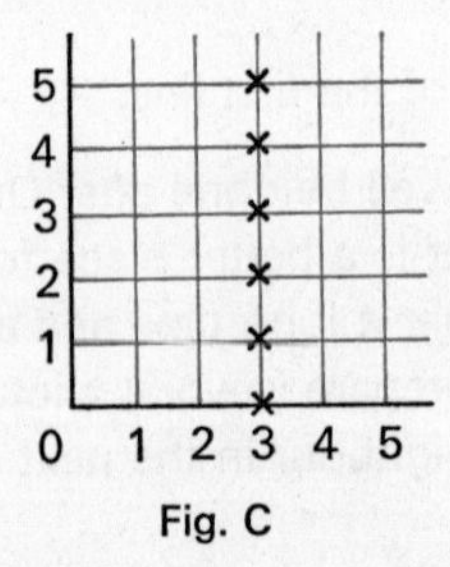

Fig. C

Then someone will suggest points in between the crosses and some of the halves can be marked. After some more thought, someone else might suggest all the quarters and so on until a pupil, tired of waiting, will come up and join all the crosses to make a line.

4 Plot on squared paper the set of points with the following coordinates. Join them up in the order in which they are written. What pictures do you get?

(*a*) (2, 1), (6, 1), (6, 3), (7, 3), (4, 6), (1, 3), (2, 3);
(*b*) (1, 1), (5, 1), (5, 6), (4, 7), (4, 11), (3, 13), (2, 11), (2, 7), (1, 6);
(*c*) (1, 1), (3, 1), (4, 3), (5, 3), (5, 5), (6, 5), (5, 7), (5, 9), (1, 9).

In (*c*), what should be put round the point (4, 7)?

5 Plot on squared paper the sets of points with the following coordinates. Which capital letters of the alphabet do they suggest?

(*a*) (1, 1), (2, 3), (3, 5), (4, 3), (5, 1), (3, 3);
(*b*) (4, 0), (4, 2), (2, 4), (6, 4);
(*c*) (1, 4), (2, 2), (3, 0), (4, 2), (5, 4);
(*d*) (1, 4), (1, 3), (1, 2), (1, 1), (2, 1), (3, 1);
(*e*) (1, 1), (2, 1), (3, 2), (3, 3), (2, 4), (1, 4), (1, $2\frac{1}{2}$).

6 Plot on squared paper the following pairs of points and join up each pair with a straight line. What are the coordinates of the middle point of each of these lines?

(*a*) (1, 4), (1, 6); (*b*) (5, 2), (1, 2); (*c*) (2, 3), (6, 5);
(*d*) (2, 5), (0, 0); (*e*) (6, 0), (0, 2); (*f*) (7, 3), (2, 7);
(*g*) (7, 6), (4, 1); (*h*) (1, 2), (3, 8); (*i*) (6, 3), (2, 4).

Look at the coordinates of the end-points of each line and the coordinates of the middle point of that line. Can you find any connection between the two? Can you now give a rule for finding the coordinates of the middle point of a line?

7 Draw the line joining (0, 6) and (6, 4) and also the line joining (4, 7) and (1, 1). Where do they meet?

8 Make up a picture that can easily be described by the coordinates of its corners. Read out the coordinates to your neighbour. See if he can plot them correctly and discover what you have drawn. Now it is your turn to discover his picture.

4. LINES

Look at Figure 8. What do you notice about the points *I*, *J* and *F*? What do you notice about their first names? Are there any other points whose first names are also 3? Have you got the point half-way between *I* and *J*? Are you sure you have *all* the points whose first names are 3?

In mathematics we call the first name (the first coordinate) of a point, the x-coordinate. We can call the whole line

$$x = 3.$$

As this expression has an equals sign, we call this the *equation* of the line.

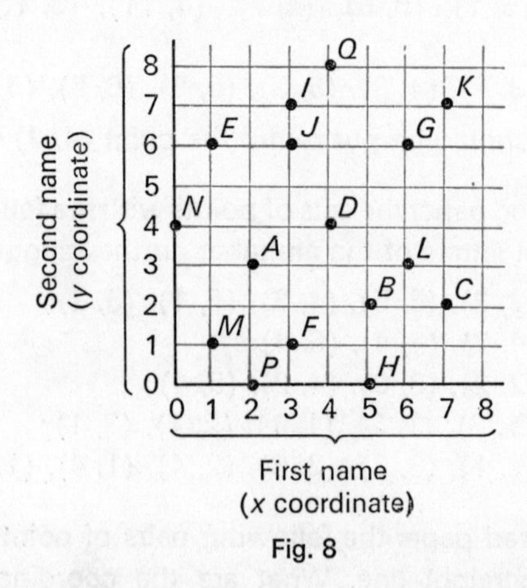

Fig. 8

Look at the points B and H. What are the x-coordinates of B and H? What are the x-coordinates of all the other points on the same line? Give the 'equation', that is the mathematical description, of the line.

Look at the points E, J and G. What do you notice about their first names? What do you notice about their second names? In mathematics we call their second names their y-coordinates. Are there other points which have the same y-coordinate? As the y-coordinate of all the points of this line is 6, the mathematical name or equation of this line is

$$y = 6.$$

Look at the points B and C. What are the y-coordinates of B and C and all the other points on the same line? What is the equation of this line?

Exercise C

What are the equations of the lines on which the following points in Figure 8 lie?

1 L and G.

2 M and E.

3 N and D.

4 I and K.

5 P and A.

6 A and L.

7 K and C.

8 Q and D.

9 M and F.

10 P and H.

Equation of line through B and H is $x=5$.
Equation of line through B and C is $y=2$.

Exercise C

1 $x=6$.

2 $x=1$.

3 $y=4$.

4 $y=7$.

5 $x=2$.

6 $y=3$.

7 $x=7$.

8 $x=4$.

9 $y=1$.

10 $y=0$.

Coordinates

The coordinates of the origin are (0, 0).
The coordinates of *A* are (3, 2).
The order of (2, 3) is wrong for point *A*.
An 'ordered pair' means 'two things taken in order'.

In order to mark the point (9, 3), all that is necessary is to draw further grid lines. For points $(2, 4\frac{1}{2})$ and $(1\frac{3}{4}, 5)$, the grid lines $y = 4\frac{1}{2}$, $x = 1\frac{3}{4}$ would be imagined.

Obviously any point can, in theory, be marked with its own pair of coordinates, even though only the numbers 1 to 6 are actually used for this grid.

Exercise D

1 (*a*) (3, 2); (*b*) (5, 6); (*c*) (4, 4);
(*d*) (2, 1); (*e*) none; (*f*) none;
(*g*) (1, 6); (*h*) (5, 0); (*i*) (2, 7);
(*j*) (2, 4).

The lines $x = 0$ and $y = 0$ are called the 'axes'. The point O, where they intersect, is called the 'origin'. What are the coordinates of the origin? See Figure 9.

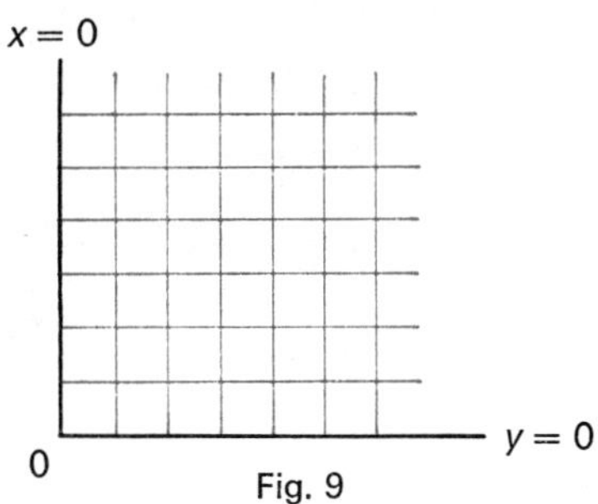

Fig. 9

This figure shows the axes with the coordinates marked along them, also the lines $x = 3$ and $y = 2$ which intersect at A. What are the coordinates of A?

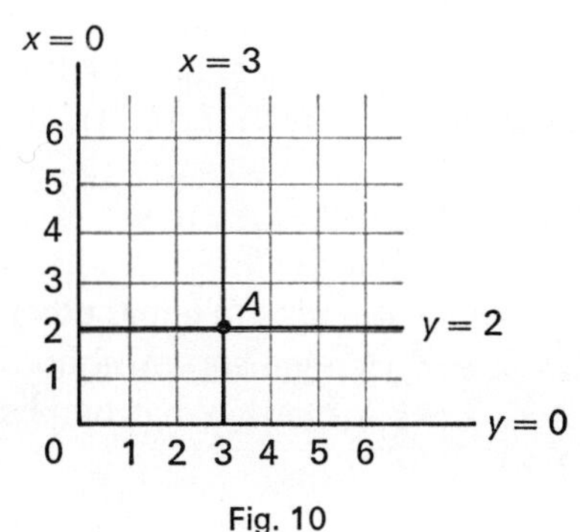

Fig. 10

The coordinates of A are given by the ordered pair (3, 2). Will the ordered pair (2, 3) give the point at A? What do we mean by 'ordered pair'?

Is it difficult to mark the point R in Figure 10 at (2, 5)?. This is easy to do, but what about the points (9, 3) or $(2, 4\frac{1}{2})$ or $(1\frac{3}{4}, 5)$?

Can we mark in any point we like?

Exercise D

1 The lines $x = 2$ and $y = 3$ intersect at the point (2, 3). At what points do the following lines intersect?

(a) $x = 3, y = 2$; (b) $x = 5, y = 6$; (c) $x = 4, y = 4$;
(d) $y = 1, x = 2$; (e) $x = 3, x = 4$; (f) $y = 1, y = 5$;
(g) $x = 1, y = 6$; (h) $y = 0, x = 5$; (i) $x = 2, y = 7$;
(j) $x = 2, y = 4$.

2 The point (3, 1) lies on the lines $x = 3$ and $y = 1$.
On what lines do the following points lie?

(*a*) (2, 4); (*b*) (5, 3); (*c*) (0, 1); (*d*) (3, 0);
(*e*) (4, 4); (*f*) (1, 7); (*g*) (0, 0); (*h*) (6, 9);
(*i*) (7, 8); (*j*) (6, 3).

3 Look back at Figure 8. Imagine the points *K*, *G*, *D* and *M* joined up. What do you notice about the line joining them? Look at the co-ordinates of these points. What do you think the equation of this line is?

5. REGIONS

On squared paper, mark with a dot each member of the set of points *A* (2, 5), *B* (1, 3), *C* (0, 6), *D* (1, 7), *E* (0, 2), *F* (2, 1). On the same paper mark with a cross each member of the set of points *G* (4, 6), *H* (6, 0), *I* (4, 1), *J* (5, 3), *K* (7, 5), *L* (6, 7).

Draw in the line $x = 3$. What do you notice about the two sets of points?

Where could you put more dots to show other points belonging to the first set? Does it matter what numbers you choose for their 'x-coordinates' (their first names)? Does it matter what numbers you choose for their 'y-coordinates'?

Where could you put more crosses to show other points belonging to the second set? Does it matter what you choose for their x-coordinates? Does it matter what you choose for their y-coordinates?

Where do you think you could NOT put any dots or crosses?

The line $x = 3$ separates the plane into two sets of points called *regions.* Are there any points that do not belong to either region? What about the points actually on the line itself?

So there are really three sets:
(i) points on one side of the line,
(ii) points on the other side of the line,
(iii) points on the line.

For the region containing the set of dots, you found that, while the y-coordinates could be anything you liked, all the x-coordinates were less

2 (*a*) $x=2, y=4$; (*b*) $x=5, y=3$;
(*c*) $x=0, y=1$; (*d*) $x=3, y=0$;
(*e*) $x=4, y=4$; (*f*) $x=1, y=7$;
(*g*) $x=0, y=0$; (*h*) $x=6, y=9$;
(*i*) $x=7, y=8$; (*j*) $x=6, y=3$.

3 Any description such as 'first name is the same as second name' would do. This could then be formalized into $x=y$.

5. REGIONS

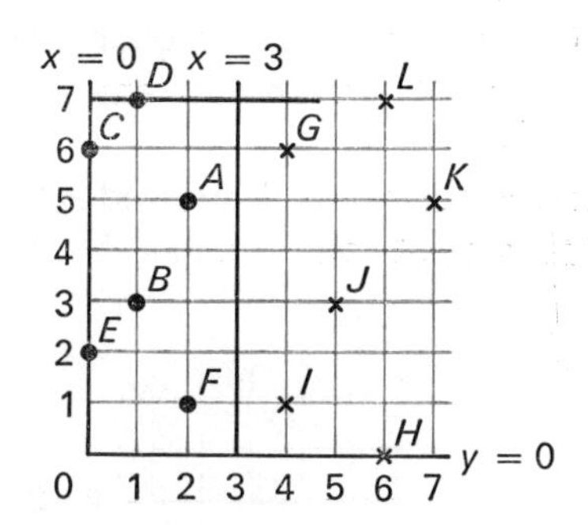

Fig. D

The regions illustrated in this section are continuous parts of the plane. Once again, it is advisable to build up the points of the region (for instance, $y>3$) first of all by letting x and y be whole numbers, then rationals and then real numbers, subject to the condition $y>3$. The extension of the region to negative numbers will be discussed in a later book. (In this chapter, the region $x<2$ is to be taken as $0 \leqslant x<2$.)

Notice that, though we are describing the lines and regions as sets of points, we are not using set notation. We are simply using the equations and orderings as the names of the lines or regions. For example, we refer to the region $x<7$ rather than $\{(x, y): x<7\}$. This notation might be introduced to the class if the chapter is going well, but it is certainly not necessary at this stage and will be mentioned in a later chapter.

If pupils wished to include the line $x=3$ with the region $x<3$, then the sign to use would be $x \leqslant 3$. Similarly, the line $x=3$ together with the region $x>3$ would be written as $x \geqslant 3$.

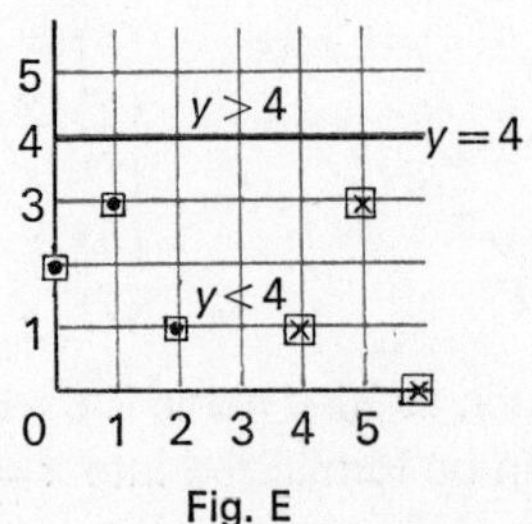

Fig. E

It is always true that, when we are asked to illustrate a region, for example, $y < 4$, we must first draw the boundary line $y = 4$.

In Figure 11, the points are on different sides of the line $x = 2$. The shaded region is $x > 2$.

No convention of shading will be stressed at this stage. We shall not say 'Indicate the region $y < 1$ on a graph' but rather 'Shade the region $y < 1$'.

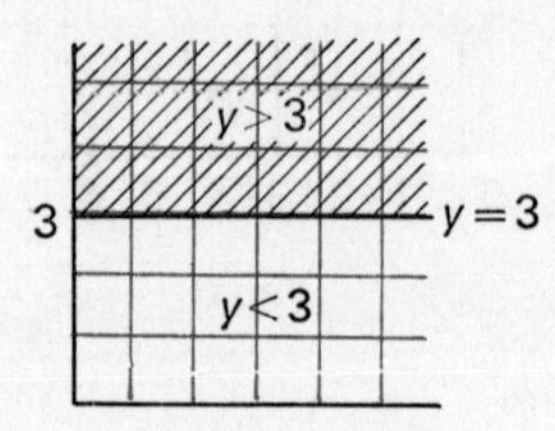

Fig. F

than 3. This is written for short as '$x < 3$'. This is the mathematical description of the region.

For the region containing the set of crosses, you found that while again the y-coordinates could be anything you liked, all the x-coordinates were bigger than 3. This is written for short as '$x > 3$'.

Look at the points A to L again. Find all the points marked whose y-coordinates are bigger than 4. Put a circle round each of them. What is the mathematical description of the region which contains these points?

Can you find all the points marked whose y-coordinates are less than 4? Put a square round each of them. What is the description of the region that contains these points?

What is the equation of the line which separates the two regions?

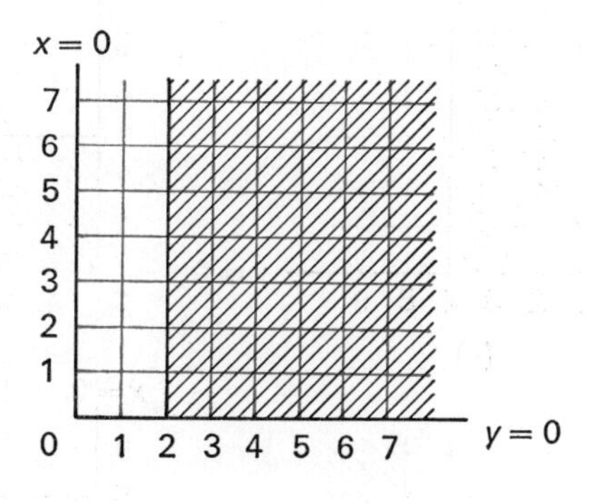

Fig. 11

In Figure 11 the points have been divided into two regions by a line. What is this line called? Look at the shaded region. Look at the x numbers marked along the bottom in this region. Is this the region $x < 2$ or $x > 2$? Look at the x numbers marked along the bottom in the unshaded region. Is this the region $x < 2$ or $x > 2$?

Sketch Figure 11 and mark the line $y = 3$ with a thick line. By looking at the y numbers along the side, decide which is the region $y > 3$ and then shade it in. What is the other region called?

Exercise E

1 Name the shaded regions in Figure 12:

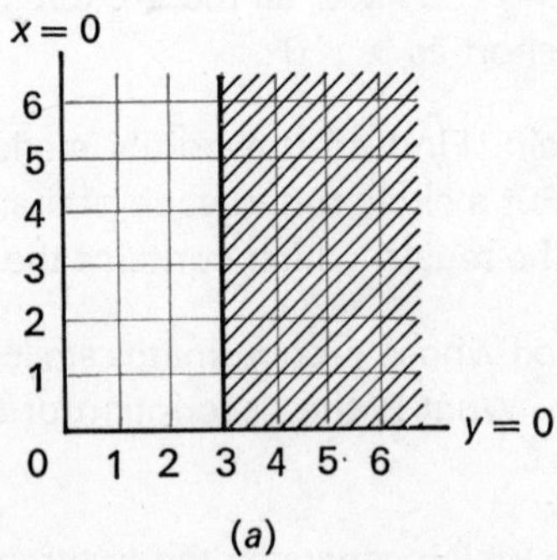

(*a*)

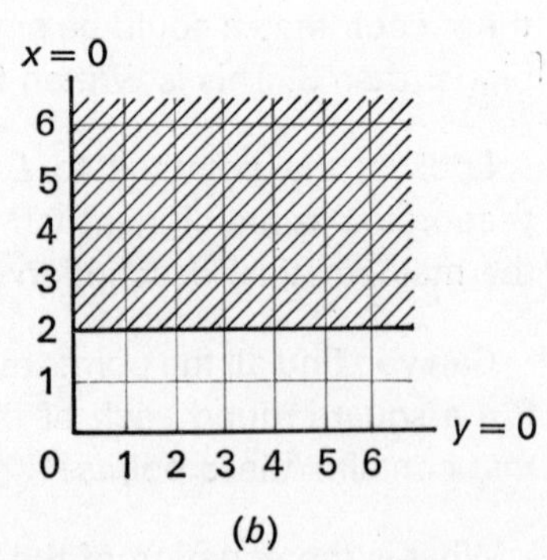

(*b*)

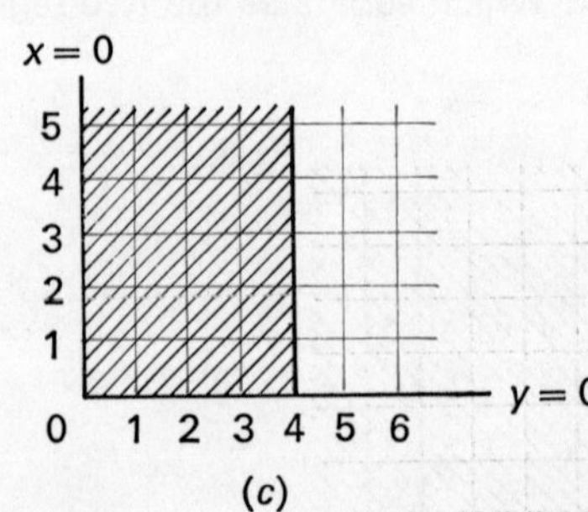

(*c*)

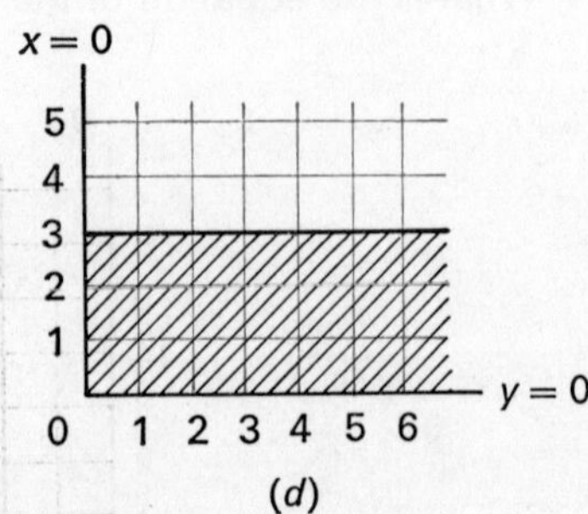

(*d*)

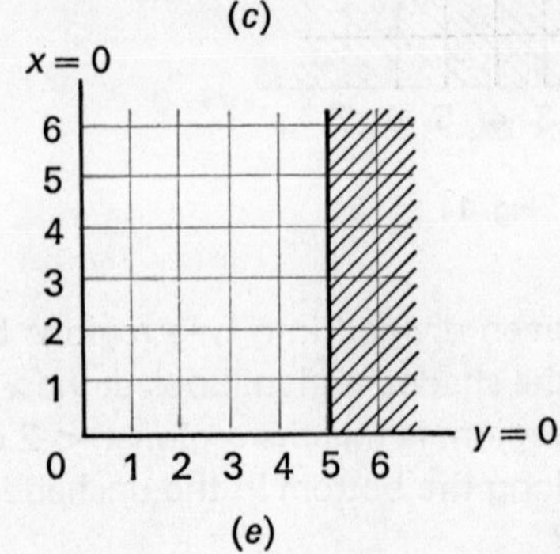

(*e*)

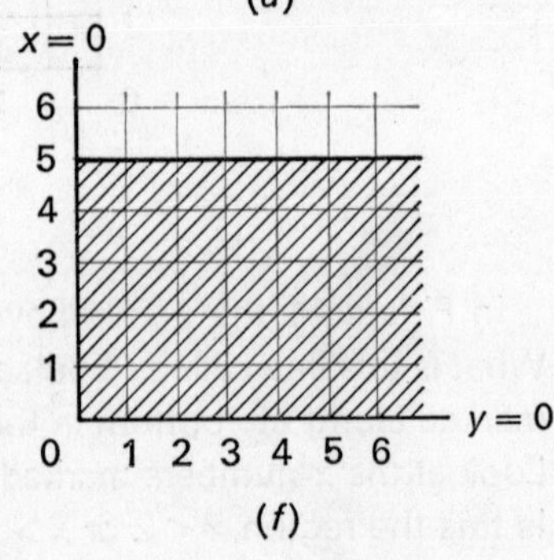

(*f*)

Fig. 12

2 By sketching diagrams similar to Figure 12, show the regions:

(*a*) $x>4$; (*b*) $y>1$; (*c*) $x<3$; (*d*) $y<2$;

(*e*) $y>4$; (*f*) $x<1$; (*g*) $x>1$; (*h*) $y<4$;

(*i*) $x>0$; (*j*) $y>0$; (*k*) $3<x$; (*l*) $5>y$;

(*m*) $4<y$; (*n*) $5>x$; (*o*) $2<x$.

Exercise E

1 (*a*) $x>3$; (*b*) $y>2$; (*c*) $x<4$;
(*d*) $y<3$; (*e*) $x>5$; (*f*) $y<5$.

2 In parts (*k*) to (*o*), the inequalities have been written the other way round to prepare the pupils for later inequalities such as $3<x<7$.

3 m is $x<3$; n is $y<4$.

(Using the set notation mentioned above, the region is given by

$$\{m \cap n\} = \{(x, y) : x<3, y<4\},$$

but again its use is not necessary at this stage.)

This question brings in for the first time the idea of the intersection of two sets P and Q, i.e. $P \cap Q$, as a simple overlap of two regions. Some further examples at this stage may possibly be found necessary, but see Chapter 8.

4

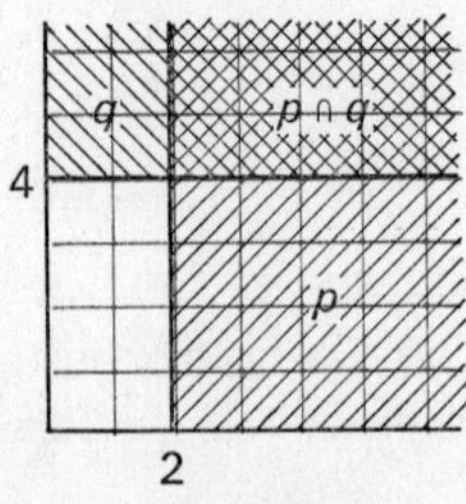

Fig. G

5

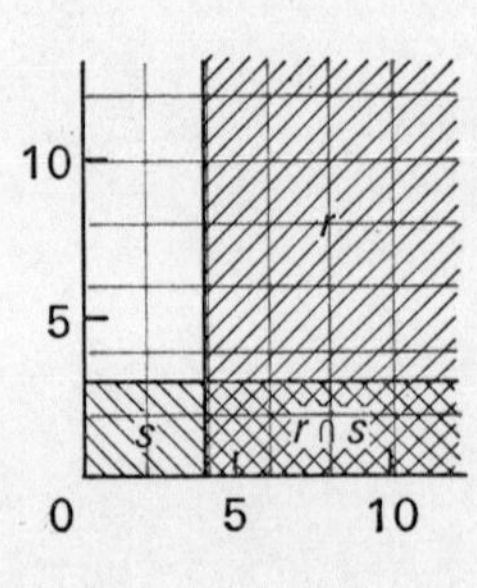

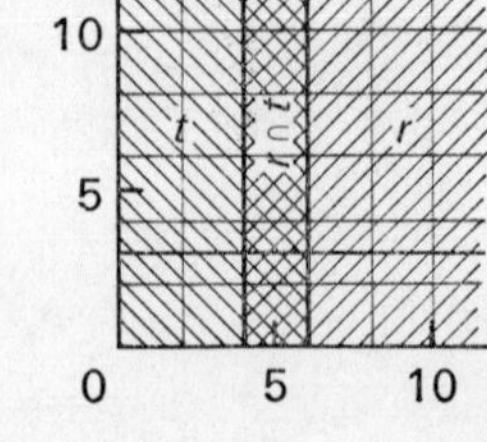

Fig. H

3

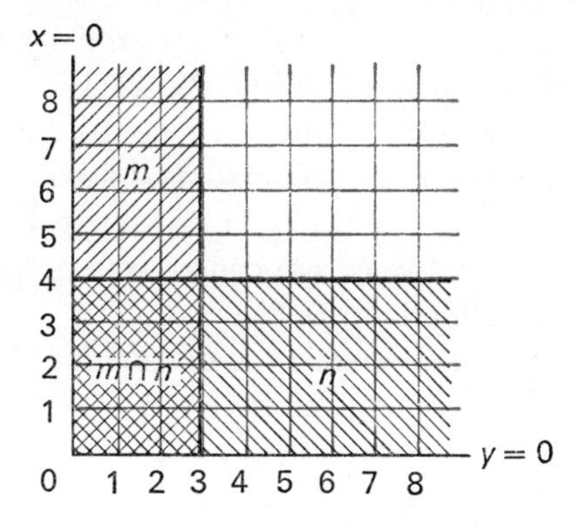

Fig. 13

Figure 13 shows two regions m and n. What are they? The overlap of the two regions has a special name. It is called 'm intersection n' and is written as '$m \cap n$'.

4 Let p be the region $x > 2$. On a diagram shade in the region p. Let q be the region $y > 4$. On the same diagram shade differently the region q. Mark in the region $p \cap q$ (that is where p and q overlap).

5 Let r be the region $x > 4$, s the region $y < 3$ and t the region $x < 6$.
(*a*) On the same diagram, show, by different shadings, the regions r and s. Mark in the region $r \cap s$.
(*b*) On another diagram, show, by different shadings, the regions s and t. Mark in the region $s \cap t$.
(*c*) Do the same for r and t.

SUMMARY

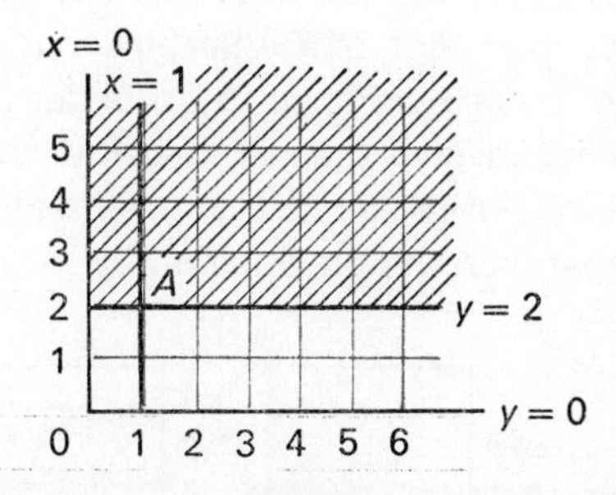

Fig. 14

We arrange the axes so that the x-coordinate of a point is taken from the set of numbers across the page and the y-coordinate from the set of numbers up the page.

Coordinates

The point A has its x-coordinate 1 and its y-coordinate 2. So its coordinates are the ordered pair (1, 2).

All points whose x-coordinate is 1 lie on the line whose equation is $x=1$. All points whose y-coordinate is 2 lie on the line whose equation is $y=2$. The lines $x=1$ and $y=2$ intersect at A, the point (1, 2).

The lines $x=0$ and $y=0$ are called axes or base lines of the system. They intersect at the origin (0, 0).

The set of points on Figure 14 is divided into three:

(i) the shaded region, given by $y>2$;

(ii) the line, whose equation is $y=2$;

(iii) the unshaded region, given by $y<2$.

Miscellaneous Exercise F

1 Plot the following sets of points. What capital letters of the alphabet do they suggest?

(*a*) (1, 4), (1, 2), (1, 0), (2, 2), (3, 2), (4, 4), (4, 2), (4, 0);

(*b*) (1, 4), ($2\frac{1}{2}$, 4), (4, 4), ($2\frac{1}{2}$, 3), ($2\frac{1}{2}$, 2), ($2\frac{1}{2}$, 1).

2 (*a*) At what point do the lines $x=4$ and $y=5$ intersect?

(*b*) What lines intersect at the point (3, 3)?

3 Plot the points (3, 2) and (3, 5). What is the equation of the line on which they lie? Shade in the region $x>3$. What is the unshaded region called?

4 Let r be the region $x>1$ and s the region $y>3$. Show by shading the region $r \cap s$.

5 See if you can play the game of Submarines. This game is slightly more difficult than that of Battleships but very similar to it. With submarines of course we have to go under the water, so we need a third coordinate to give the depth below the surface. The game is played on 4 layers of 16 squares each, as in Figure 15, each layer being one unit deeper than the previous one.

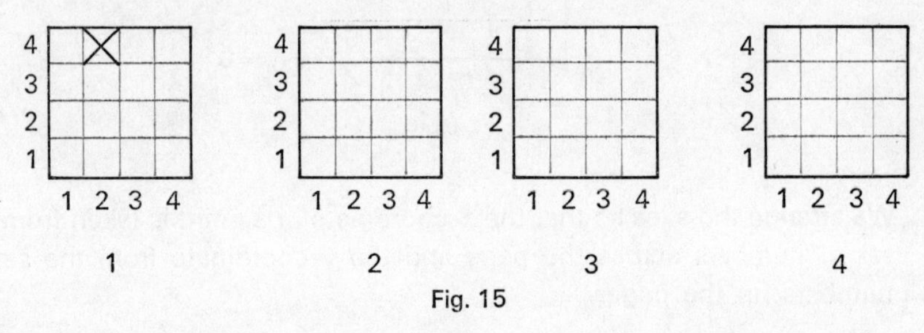

Fig. 15

Miscellaneous Exercise F

1

(a)

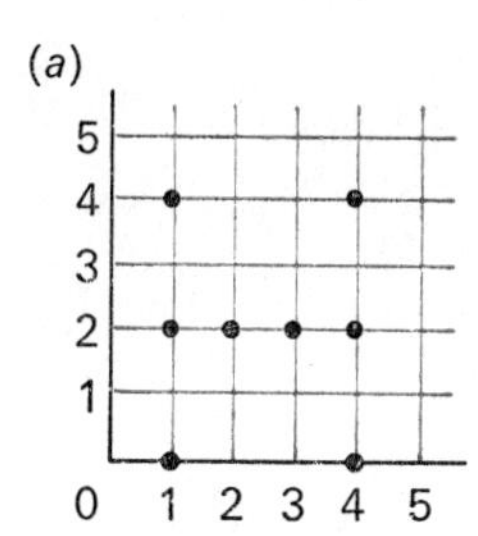

(b)

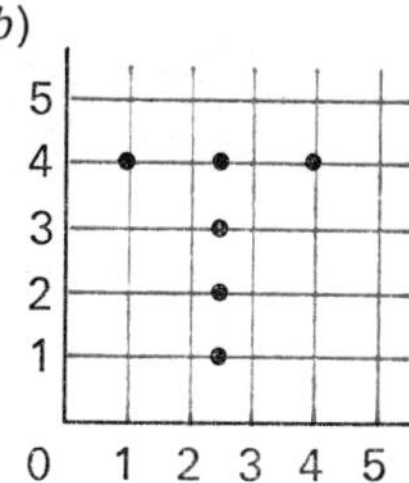

Fig. I

2 (a) (4, 5); (b) $x=3$, $y=3$.

3

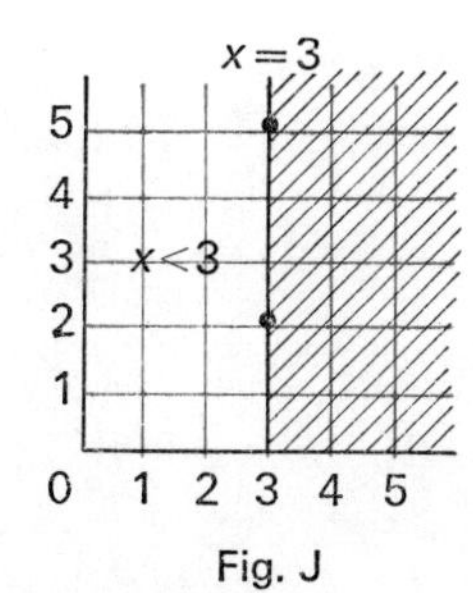

Fig. J

4

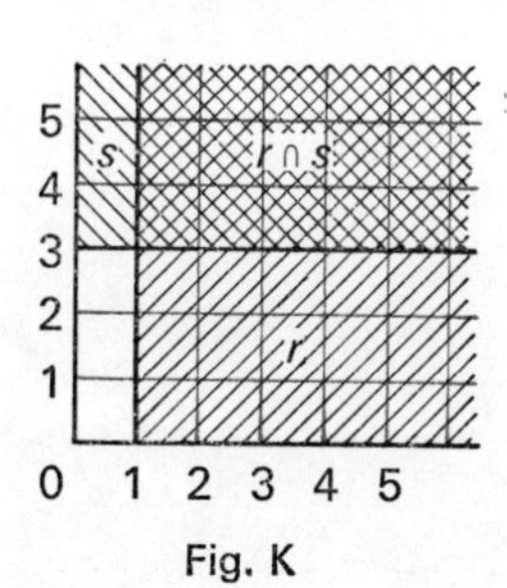

Fig. K

5 and 6 These questions are for those members of the class who might like to try their hands at work in three dimensions.

Each fleet is as follows:

1 polaris submarine

3 killer submarines

5 midget submarines

The game is now played exactly like Battleships, each player taking it in turn to fire depth charges. A cross has been put to show a depth charge exploding at (2, 4, 1).

6 You will have played noughts and crosses on many occasions and may have exhausted all the possibilities, knowing how to win or how not to lose! We can extend the game into three dimensions.

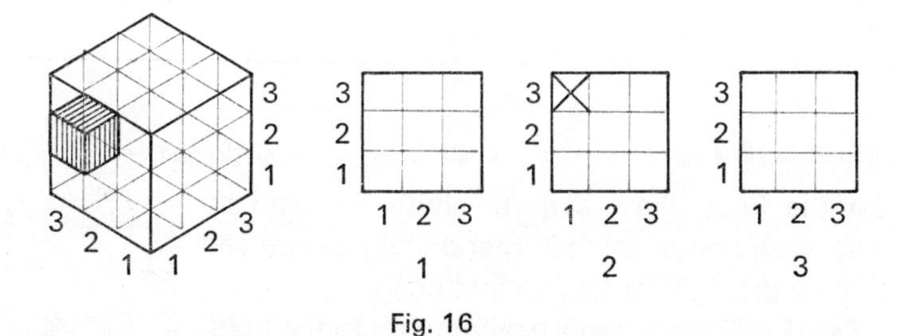

Fig. 16

Instead of 9 squares we have 27 squares such that each set of 9 squares is thought of as a layer: a bottom layer, a middle layer and a top layer. (Think of a building with 9 rooms on the ground floor, 9 on the first floor and 9 on the top floor.) The × is shown in the position (1, 3, 2). The object of the game is to obtain not only one set of three ×'s or ○'s in a line but as many as possible. Examples of winning lines are

(3, 3, 1), (3, 3, 2), (3, 3, 3);
(1, 3, 1), (1, 2, 2), (1, 1, 3);
(3, 1, 1), (2, 2, 2), (1, 3, 3).

3. Angles

Experiment 1

On the floor, draw a north–south line and an east–west line crossing it. Stand at the centre and turn to the right or left, as instructed.

(*a*) Face north, make a whole turn to the right, three whole turns to the left, two whole turns to the right. In what direction were you facing after each set of turns?

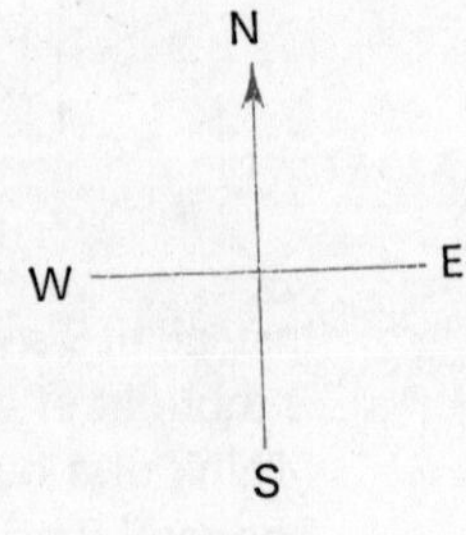

Fig. 1

(*b*) Notice in which direction you are facing at the end of each movement.

(i) Face east, make a half-turn to the left.
(ii) Face north, make a quarter-turn to the right.
(iii) Face south, make a three-quarter turn to the left.
(iv) Face west, make one-and-a-half turns to the left.

Would you be facing in the same direction after each movement if you had turned the opposite way by mistake?

(*c*) Draw the lines as before but add the north–east—south–west and the north–west—south–east lines.

You make a one-eighth turn in turning from any one line to the next. In what direction would you be facing after each of the following turns?

Face east, make one-eighth turn left.

Face south–west, make one-half turn to the right.

Face south, make five-eighths of a turn to the left.

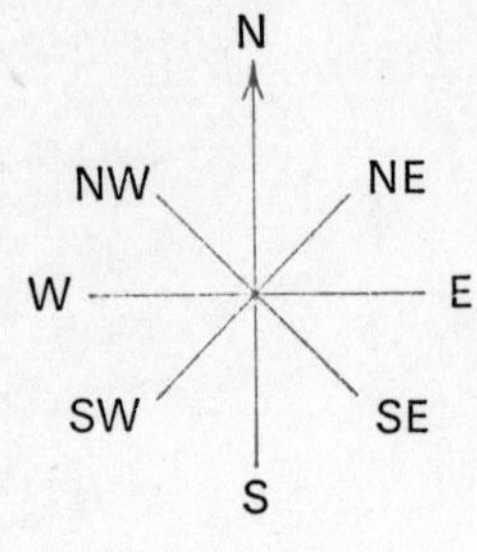

Fig. 2

3. Angles

For the whole of this course, we shall emphasize the idea of angle as an amount of rotation in much the same way that distance may be considered as an amount of translation. This is done primarily because we believe that this dynamic view of angle leads to the consideration of more interesting situations than the static view, and because it is an idea very easily grasped by pupils. It also points in the direction of the approach to geometry which will occur in later books.

Two static definitions of angle are:

the intersection of two closed half-planes whose straight line boundaries are distinct;

a pair of half-lines with a common origin.

Both these definitions are used in rigorous axiomatic developments of Euclidean geometry.

Occasionally we shall use phrases which point towards the second of these static definitions rather than the rotation definition. When we do this, we shall do it for the sake of brevity. For example, we shall refer to 'acute angles' or 'right-angles' and possibly seem to suggest a picture of a pointed or a 'square' corner rather than a measure of rotation. We shall say, 'Draw an angle of 30°' (though it is not possible to draw an amount of rotation) when we are really asking the pupils to draw the beginnings of two half-lines from a point so that a rotation of 30° would carry one onto the other.

Given a pair of half-lines, there are two possible rotations transforming one onto the other; so the notion of angle as a pair of half-lines is less precise than that of angle as a rotation. Strictly speaking, one should consider an ordered pair of half-lines together with a positive (anti-clockwise) and a negative (clockwise) rotation about their intersection which transfers one into the other.

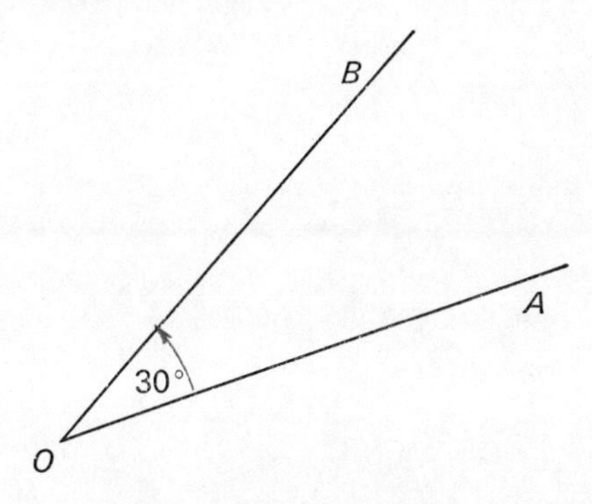

Thus the angle (OA, OB) is equal to 30°, whilst the angle (OB, OA) is equal to 330° or to ⁻30°. Traditional usage is ambiguous but we have thought it preferable to risk ambiguity rather than to adhere rigidly to the rotation concept.

We have used the curved arrow to accentuate the idea of rotation and it certainly helps when angles are measured by a protractor. In later chapters, spatial symmetries may be seen in terms of rotation, while a rotating unit vector will be considered during the introduction of sines and cosines of the general angle.

Experiment 1

It was thought necessary to introduce the idea of rotation in a practical way by suggesting that pupils themselves make the rotations, describing them in terms of their directions before and after.

(*a*) North.

(*b*) (i) West; (ii) east; (iii) west; (iv) east.

Only in (i) and (iv). In (ii) you would be facing west and in (iii) you would be facing east.

(*c*) North-east; north-east; north-west.

1. ROTATIONS

(*b*) The difference lies in the direction of turn not in the centre of rotation or the amount of rotation.

It is unfortunate that we have to use the word 'turn' both to indicate rotation, a movement, and also to indicate an amount of turn 'a quarter-turn'. (So we have inelegant questions such as 'In what direction are you facing after you have turned through a quarter-turn?'.)

1. ROTATIONS

We often turn things or parts of things. Already today you will probably have

turned a door knob to open a door;
turned a door on its hinges;
turned a water tap on and off;
screwed and unscrewed the cap of a toothpaste tube;
turned your head to look at the person beside you;
turned your whole body when moving about the house.

(*a*) What other things can you think of that turn, rotate, revolve or spin?

(*b*) What is the difference between the movements needed to turn the tap on and to turn it off?

When we rotate things we sometimes have to be careful how much we rotate them. In adjusting the hands of a watch we have to be quite exact in the amount we twist the winder. Many things have marks on them to help us to judge the amount of a turn; the numbers on the knobs of a cooker or television set or radio, for example. Name some other objects on which dials or knobs are marked to show an amount of turn.

A *rotation* is the mathematical name for a turn. A rotation has a centre of rotation, a direction of rotation and a size of rotation.

The point marked with a red dot is the centre of rotation of the illustrations in Figures 3, 4 and 5. In the examples of the door knob, toothpaste tube or when you turn, there is always a hub or axis at the centre of the turn.

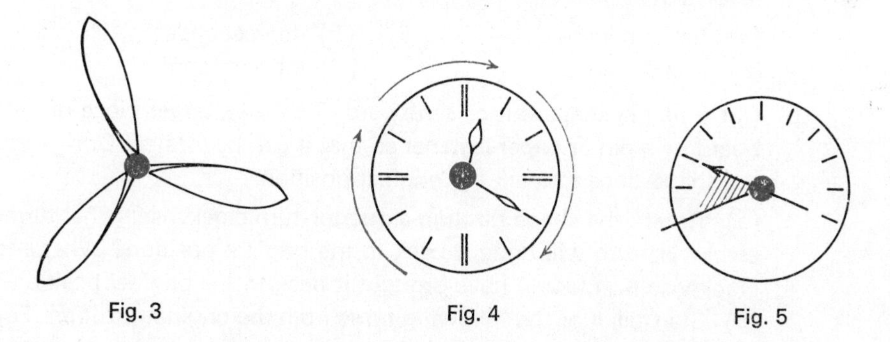

Fig. 3 Fig. 4 Fig. 5

The direction of rotation is indicated with reference to a clock (see Figure 4). It can be clockwise or anticlockwise.

For the time being, we shall use a whole-turn as the basic unit for measuring the size of a rotation. The needle of the dial of Figure 5 has moved through an eighth turn.

Exercise A

1 Experiment to discover the direction of turn and its size, to the nearest half-turn, when you carry out the following actions. Give a rough description of the centre of rotation in each case.

(*a*) Unlock your front door.
(*b*) Screw or unscrew the top of the toothpaste tube.
(*c*) Turn the volume knob of the radio or T.V. set fully on.
(*d*) Dial 9 on a telephone.
(*e*) Turn the cold tap of the bath fully on (and fully off).

2 The minute hand of a clock goes through a whole turn each hour. What part of a turn does it make in:

(*a*) 2 hours; (*b*) 5 hours; (*c*) $3\frac{1}{2}$ hours; (*d*) $\frac{1}{2}$ hour;
(*e*) $\frac{1}{4}$ hour; (*f*) 20 min; (*g*) 45 min; (*h*) 5 min?

3 What time passes when:

(*a*) the minute hand rotates through a quarter-turn;
(*b*) the hour hand rotates through one whole turn;
(*c*) the hour hand rotates through a quarter-turn;
(*d*) the second hand rotates through a half-turn?

4 A sewing machine makes 3 stitches for every turn of the flywheel. How much does the flywheel turn when the machine makes:

(*a*) 6 stitches; (*b*) 15 stitches; (*c*) 1 stitch; (*d*) 2 stitches?

5 A record makes 45 revolutions per minute. How many times does it revolve in:

(*a*) half a minute; (*b*) ten seconds;
(*c*) four seconds; (*d*) one second?

6 Cut a simple shape out of cardboard. Fix it to a larger piece of cardboard by a pin or paper fastener so that it can be rotated. Draw round the shape once to mark the original position.

(*a*) Rotate the shape through a quarter-turn clockwise. What further clockwise turn will bring it back to the original position? What anticlockwise turn would have brought it back to the original position?

(*b*) Start each of the following turns from the original position. Four of them will bring you to the same new direction. Which one is the exception?

(i) A quarter-turn clockwise;
(ii) a three-quarter turn anticlockwise;
(iii) a half-turn clockwise;
(iv) two-and-a-quarter turns clockwise;
(v) one-and-three-quarter turns anticlockwise.

Exercise A

2 (*a*) 2 turns; (*b*) 5 turns; (*c*) $3\frac{1}{2}$ turns; (*d*) $\frac{1}{2}$ turn;
(*e*) $\frac{1}{4}$ turn; (*f*) $\frac{1}{3}$ turn; (*g*) $\frac{3}{4}$ turn; (*h*) $\frac{1}{12}$ turn.

3 (*a*) 15 minutes; (*b*) 12 hours; (*c*) 3 hours;
(*d*) 30 seconds.

4 (*a*) 2 turns; (*b*) 5 turns; (*c*) $\frac{1}{3}$ turn; (*d*) $\frac{2}{3}$ turn.

5 (*a*) $22\frac{1}{2}$; (*b*) $7\frac{1}{2}$; (*c*) 3; (*d*) $\frac{3}{4}$.

6 (*a*) $\frac{3}{4}$ turn; $\frac{1}{4}$ turn.
(*b*) (iii).
(*c*) The order of turning makes no difference to the final result. Rotations are said to be commutative.

2. CORNERS AND ANGLES

It should be noticed that there are two angles at each corner in Figure 7 and, in fact, in any figure containing two rays from a single point. We have thought it reasonable to suppose that pupils would naturally consider the angle of the shaded part in each case, but the idea of a pencil being rotated from one ray to the other is more complicated and one should perhaps refer to Figure 36. The four possibilities are shown in the following figure.

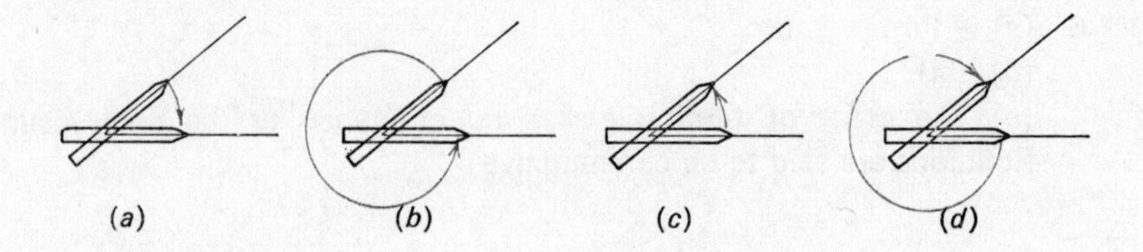

(*a*) (*b*) (*c*) (*d*)

We consider that (*c*) shows the appropriate rotation.

(*c*) Starting from the original position, make first a quarter-turn clockwise then a three-quarter turn anticlockwise. Starting again from the original position, reverse the order of these turns. Are you facing in the same direction after both experiments?

Try doing other turns first in one order and then in another. Does the order ever make any difference to the final direction?

2. CORNERS AND ANGLES

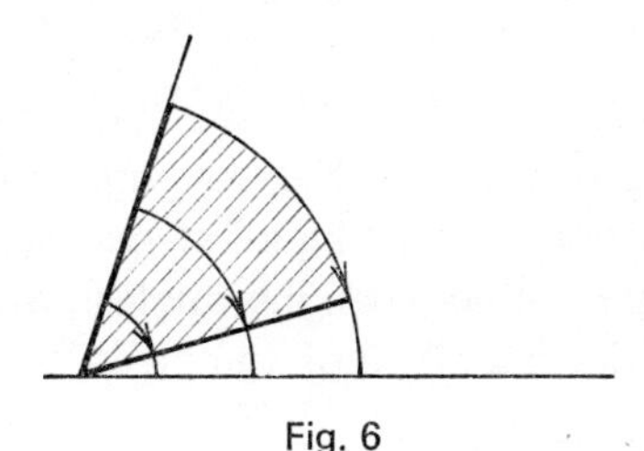

Fig. 6

An object moving along a line, moves through a *distance*. An object, rotating about a point, rotates through an *angle*. The angles we have met so far have been measured in turns or fractions of a turn. In turning from north to east, you rotate through an angle called a 'quarter-turn'.

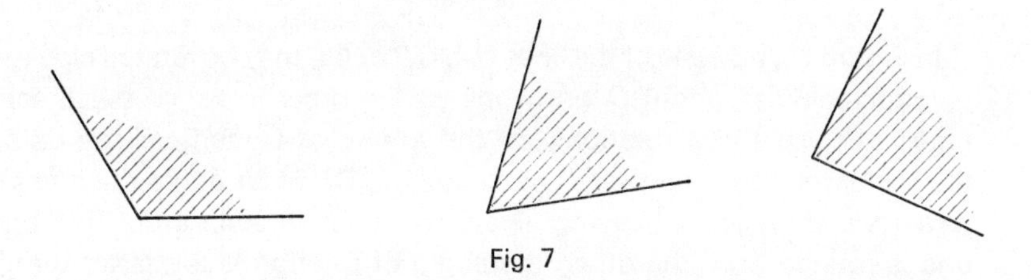

Fig. 7

One way of comparing the pointedness of these corners is by comparing the angles at the corners. A pencil set on one line of the corner can be rotated about the point of the corner until it lies on the other line. The amount of rotation is an indication of the pointedness of the corner: the bigger the rotation, the less pointed the corner.

A searchlight set in the corner between two walls can be turned to shine along one wall. Then it can be turned until it shines along the other. The amount of rotation is a measure of the angle between the walls (see Figure 8).

When cutting lino, a knife is drawn along to the point of the corner and is then turned through the angle marked in red and drawn away from the corner. The angle of the shaded piece of lino together with the angle of

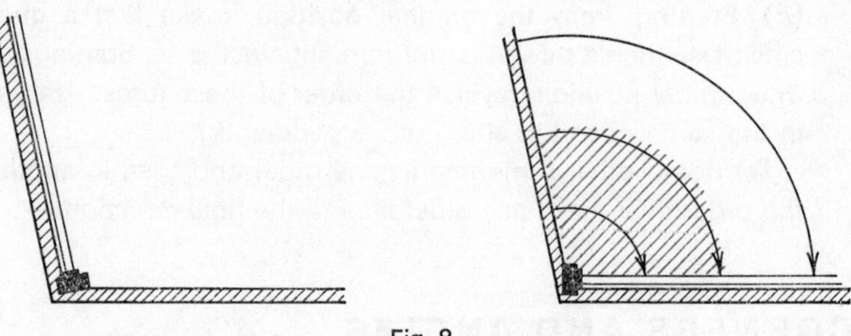

Fig. 8

the knife turn make a half-turn. When you are walking along the side of a building and then turn a corner, the amount by which you turn added to the angle of the corner of the building, makes a half-turn. When cutting out some material with scissors, the angle turned by the scissors at the corner together with the angle of the piece of material cut out, make a half-turn. See Figure 9.

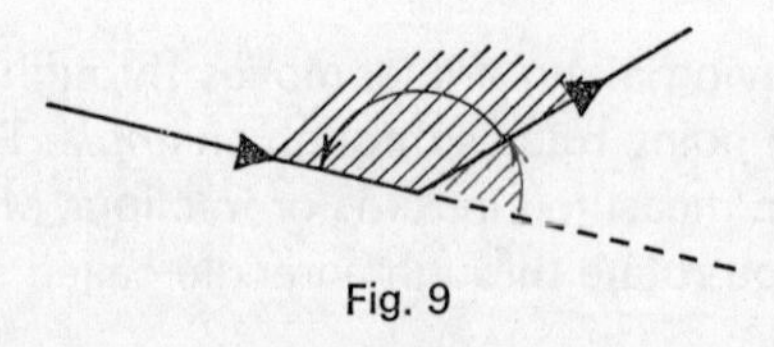

Fig. 9

In Figure 10, any one of the lines *OA*, *OB* or *OC* may be rotated clockwise or anticlockwise about *O* onto one of the other lines. We shall sometimes refer to these rotations as 'the angles at *O*' and call the point *O* the *vertex* of these angles. (The sentence 'Draw an angle of a quarter-turn,' is a short way of saying 'Draw two lines in such a position that if one is rotated onto the other, the angle of rotation is a quarter-turn.')

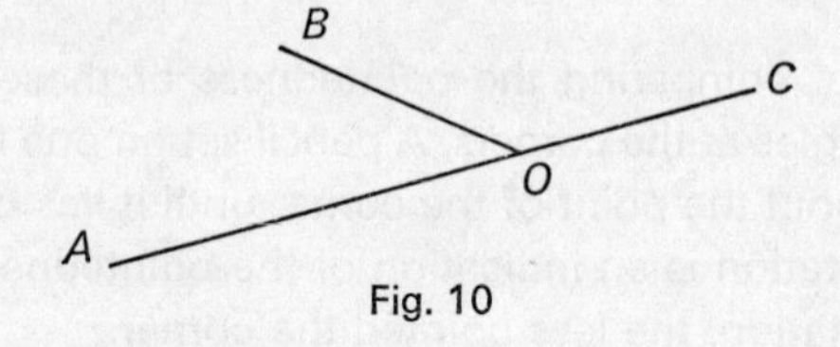

Fig. 10

To distinguish the various angles at *O*, we shall use the notation $\angle AOB$ to indicate the angle obtained by rotating *OA* onto *OB*; or $\angle BOC$ to indicate the angle obtained by rotating *OB* onto *OC*. (In each case, it is the smaller of the two possible rotations that is referred to.)

When *AOC* is a straight line we can write that

$$\angle AOB + \angle BOC = \text{a half-turn.}$$

Referring to Figure 10, it is often stated that the angles *AOB* and *BOC* are angles on the straight line *AC*, but this might be considered confusing if we are concentrating upon rotation. We therefore refer to the angles *AOB* and *BOC* at *O* and say that the sum of these angles is a half-turn. We further say that the rays *OA* and *OC associated with* the angles *AOB* and *BOC* form a straight line, provided that the sum of the angles equals a half-turn.

Exercise B

2 In both cases the final direction relative to the starting direction will be the same whether the rotation is to the left or to the right.

4 (*a*) $\frac{1}{4}$ turn; (*b*) $\frac{1}{4}$ turn; (*c*) $\frac{3}{8}$ turn;
(*d*) $\frac{3}{8}$ turn; (*e*) $\frac{3}{4}$ turn.

5 (*a*) $\frac{1}{4}$ turn clockwise; (*b*) $\frac{3}{4}$ turn anticlockwise;
(*c*) $\frac{1}{8}$ turn anticlockwise; (*d*) $\frac{1}{2}$ turn clockwise;
(*e*) whole-turn clockwise.

6

(*a*)

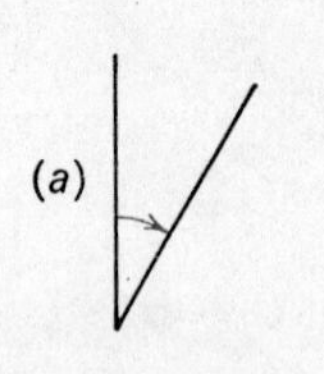

(*b*)

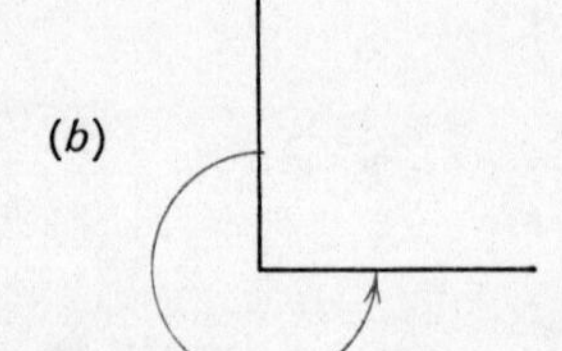

(*c*)

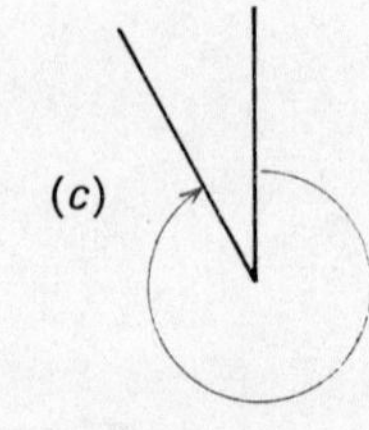

(*d*)

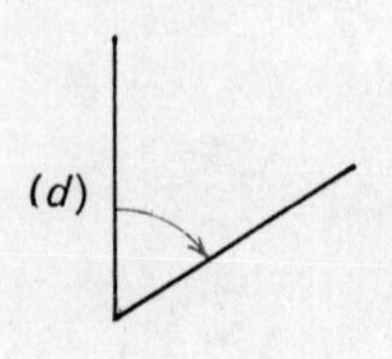

(*e*)

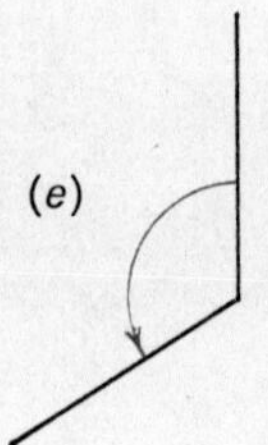

(*f*)

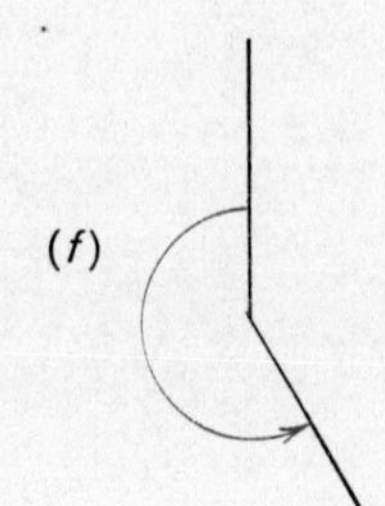

Exercise B

1 Sketch an angle for each rotation:

(a) $\frac{1}{4}$ turn left; (b) $\frac{1}{8}$ turn right; (c) $\frac{3}{4}$ turn right;
(d) $\frac{1}{6}$ turn left; (e) $\frac{5}{8}$ turn left; (f) $\frac{3}{8}$ turn right.

2 What is special about these angles:

(a) a half-turn right or left;
(b) a whole-turn right or left?

3 The corner of Figure 11 is $\frac{1}{6}$ turn. Trace it to help you draw the angles:

(a) $\frac{1}{6}$ turn right;
(b) $1\frac{1}{6}$ turn right;
(c) $\frac{1}{6}$ turn left;
(d) $2\frac{1}{6}$ turn left;
(e) $3\frac{1}{6}$ turn right.

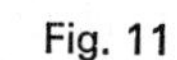

Fig. 11

4 Through what angle do you turn in going,

(a) from south clockwise to west;
(b) from north-west clockwise to north-east;
(c) from north clockwise to south-east;
(d) from north-west anticlockwise to south;
(e) from north-east anticlockwise to south-east?

5 Describe these angles:

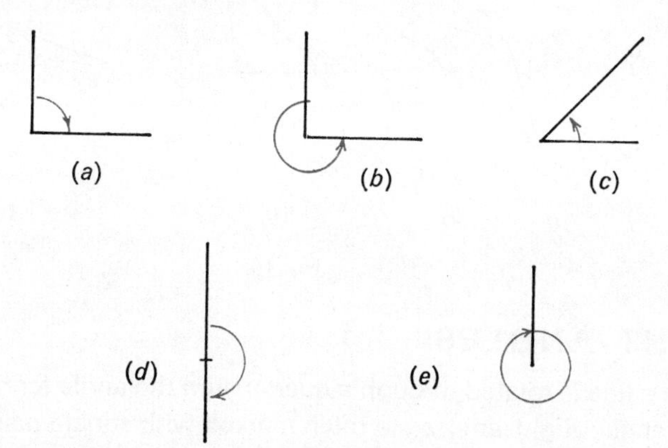

Fig. 12

6 A clock face shows twelfths of a whole-turn.
Draw these angles:

(a) $\frac{1}{12}$ turn clockwise; (b) $\frac{9}{12}$ turn anticlockwise;
(c) $\frac{11}{12}$ turn clockwise; (d) $\frac{2}{12}$ turn clockwise;
(e) $\frac{4}{12}$ turn anticlockwise; (f) $\frac{7}{12}$ turn anticlockwise.

7 These wedges of cheese have slightly different angle sizes. Give them in order from the smallest angle to the largest.

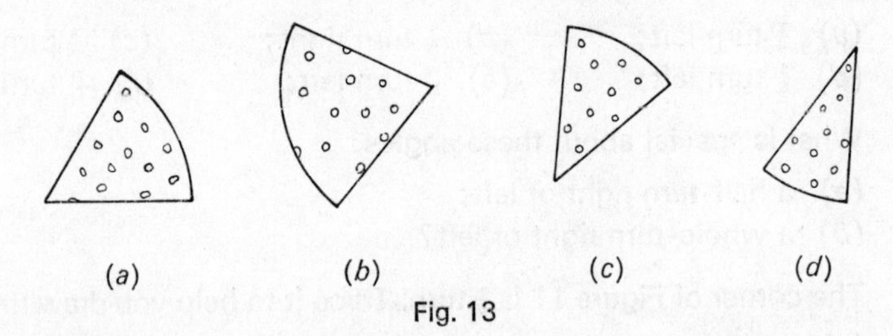

Fig. 13

8 A man is fitting lino at this corner of a floor.

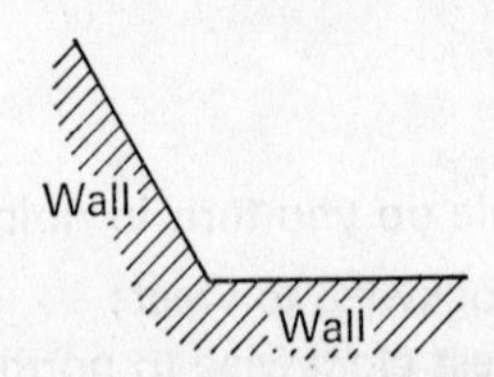

Which of these pieces would fit the corner?

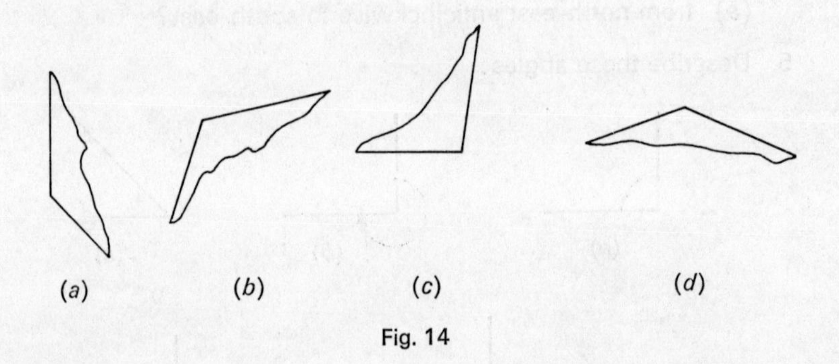

Fig. 14

3. RIGHT-ANGLES

When a line is rotated through a quarter-turn the angle formed is called a *right*-angle. Right-angles are often marked with square marks.

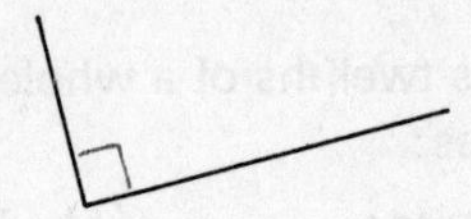

It is often unnecessary to show the direction of turn and, from now on, we shall sometimes not mark the direction of the angle.

7 (*d*), (*c*), (*a*), (*b*).

8 (*b*).

3. RIGHT-ANGLES

The name *right-angle* is given to what has so far been called a quarter-turn. You will notice that the grammar of such phrases as right-angle, acute angle or reflex angle seems to be more suited for the concept of angle as two rays from a single point rather than as a measure of rotation. Nevertheless, we think it can be extended to the dynamic definition of an angle.

T

Angles

The paper unfolded shows two lines that cross and in fact this defines four angles whether the lines are perpendicular or not. In such cases, we have to be careful in lettering angles.

There is a method of making an accurate right-angle without measuring the rotation. Take a piece of paper (or cloth or a leaf) and fold it once. Then fold it again, as in the picture, so that *A* meets *B*.

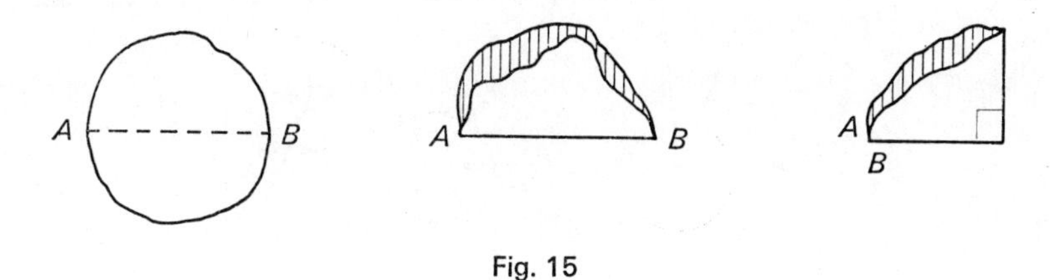

Fig. 15

You have divided a whole turn into quarters and made four right-angles. When you unfold the paper you can see that the angle about any point on a straight line is two right-angles. Two straight lines that cross at right-angles are said to be *perpendicular* to each other.

Look for some examples of right-angles. You should be able to count at least 200 of them in a day.

We can compare other angles with right-angles.

An *acute* angle is an angle less than a right-angle.

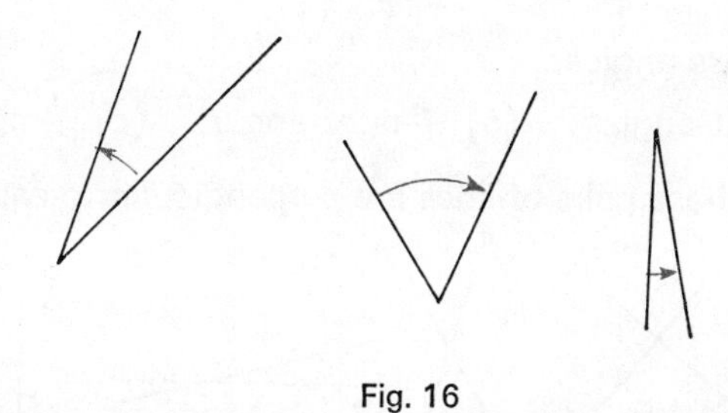

Fig. 16

An *obtuse* angle is an angle greater than one right-angle but less than two right-angles.

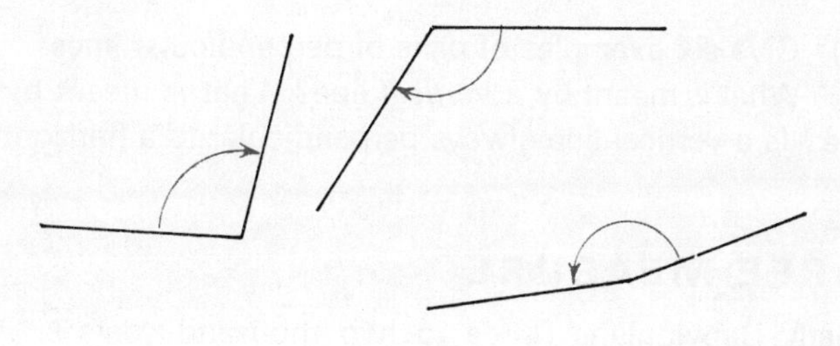

Fig. 17

A *reflex* angle is an angle greater than two right-angles but less than four right-angles (see Figure 18).

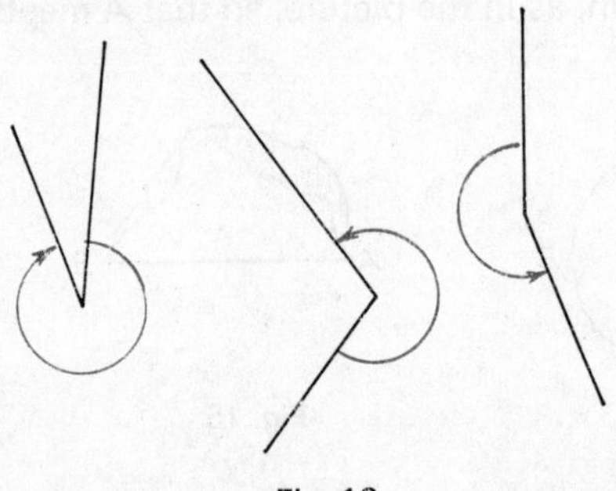

Fig. 18

Exercise C

1 Which angles are acute? Which angles are obtuse? Which angles are reflex?

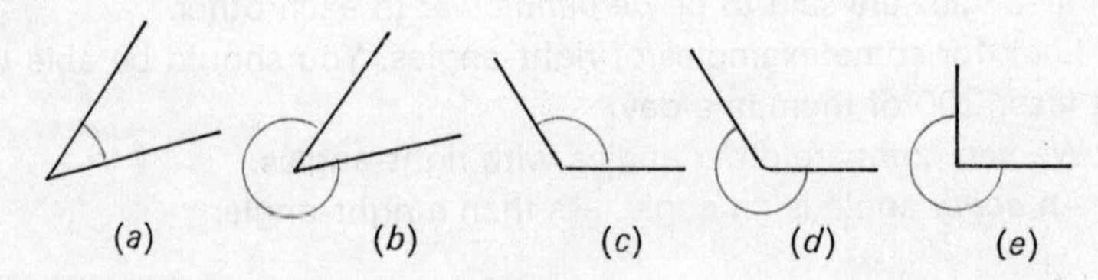

(*a*) (*b*) (*c*) (*d*) (*e*)

Fig. 19

2 Sketch these angles:

(*a*) 3 right-angles; (*b*) $\frac{1}{2}$ right-angle; (*c*) $\frac{3}{4}$ right-angle.

3 Which of these pairs of lines are perpendicular to each other?

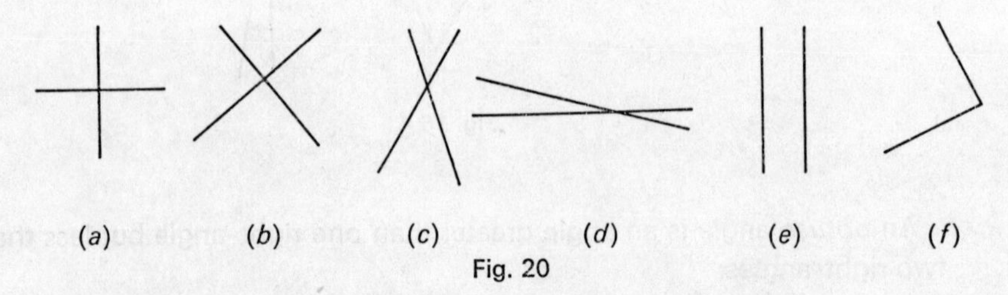

(*a*) (*b*) (*c*) (*d*) (*e*) (*f*)

Fig. 20

(*a*) Give six examples of pairs of perpendicular lines.
(*b*) What is meant by a *vertical* line? What is meant by a *horizontal* line? Is a vertical line always perpendicular to a horizontal line?

4. DEGREE MEASURE

The early Babylonians (three to two thousand years B.C.) were great astronomers, and needed accurate measurements to describe the move-

Exercise C

1 (*a*) is acute; (*c*) is obtuse;
(*b*), (*d*) and (*e*) are reflex.

2

(*a*) 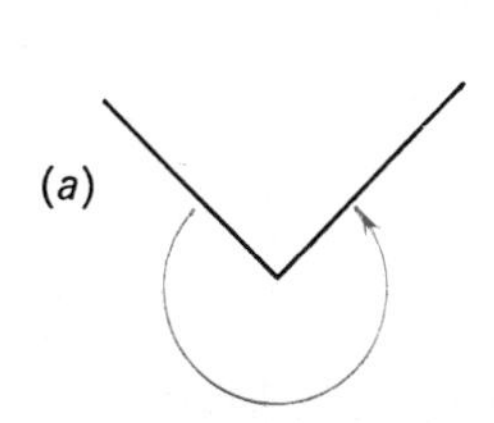 (*b*) 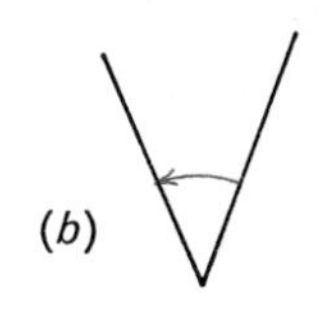 (*c*) 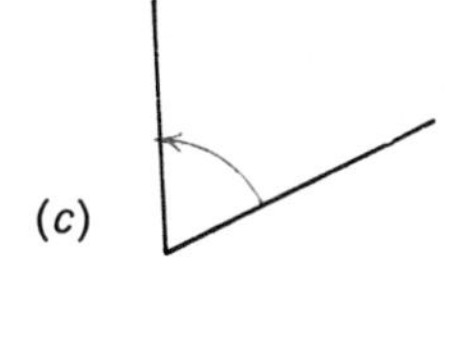

3 In (*a*), (*b*) and (*f*) the lines are perpendicular to each other.
(*b*) The words 'vertical' and 'horizontal' apply to the direction of lines relative to the earth. A vertical line at a point is one whose direction is towards the centre of the earth. This is the direction that would be taken by a plumb line at that point. Lines and planes perpendicular to the vertical line at a point are said to be horizontal.

From the above definitions, horizontal and vertical lines at a point are always perpendicular but, of course, a vertical line at one point is not perpendicular to the horizontal line at another because the earth is round.

4. DEGREE MEASURE

(*a*) 360°.
(*b*) 180°.
(*c*) 90°.
(*d*) 30°.
(*e*) 15°, 90°, 150°.

ments of the planets. They probably based their units on one-sixth of a turn, which they split into sixtieths (they were particularly interested in sixties), and called 'degrees'.

60° (read 'sixty degrees') is a sixth of a whole turn.

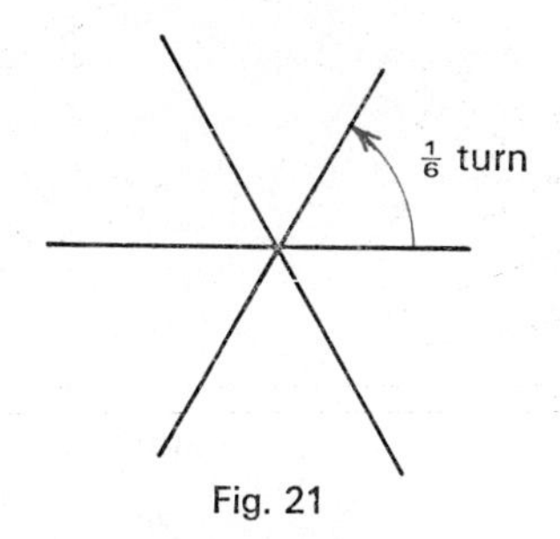

Fig. 21

(*a*) How many degrees is a whole turn?
(*b*) How many degrees is half a turn?
(*c*) How many degrees is a right-angle?
(*d*) How many degrees is a one-twelfth turn?
Here is an angle of 1 degree.

Here are some angles with their size shown in degrees.

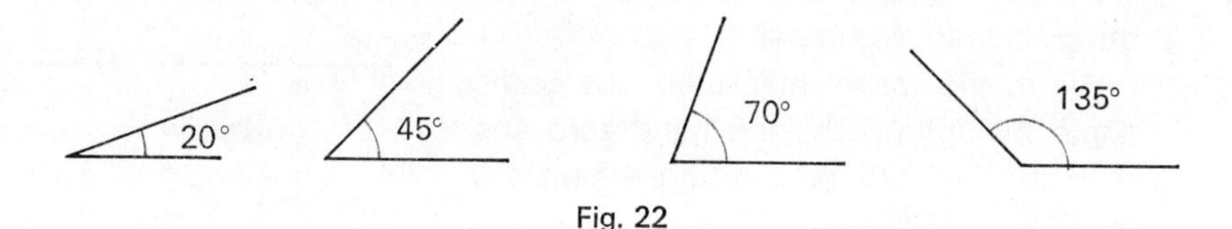

Fig. 22

(*e*) Use the above angles to estimate the size of the following ones in degrees.

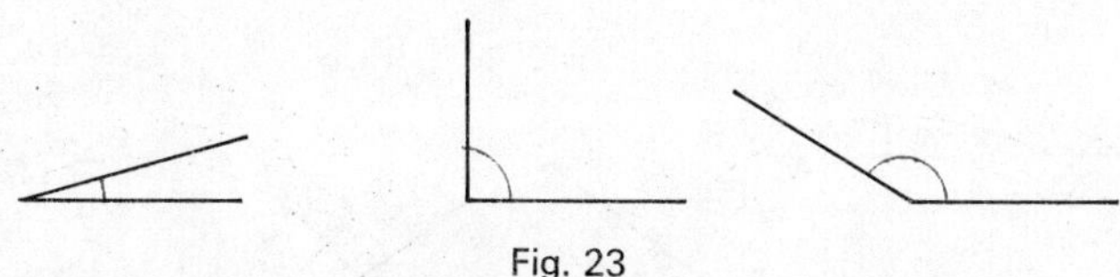

Fig. 23

(*f*) Make sketches to illustrate angles that are roughly of these sizes:

90°, 60°, 45°, 180°, 110°.

4.1 Use of protractor

To draw and measure angles accurately we need an instrument called a '*protractor*' (see Figure 24). The angle sizes are marked in degrees.

It is usually semi-circular and made of transparent plastic. The idea is to compare the angle marked on the protractor with that on the paper.

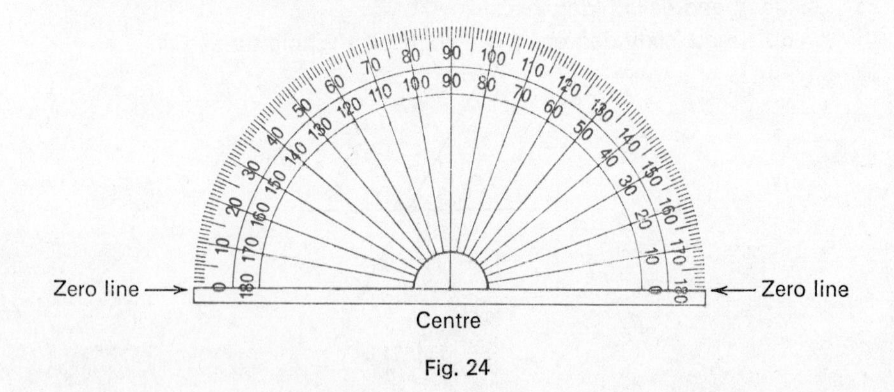

Fig. 24

This protractor has two scales. The outer scale goes 0 ↷ 180° clockwise; the inner scale goes 180° ↶ 0 anticlockwise.

(*a*) To measure the angle *BOC*, place the centre of the protractor over the vertex *O*.

Fit the right-hand zero line over the line *OC*. Note where the line *OB* cuts the inner scale—the one whose zero is on *OC*. Why would it have been wrong to use the outer scale in this case?

We could have measured the same angle by putting the left-hand zero line over the line *OB* and noting where *OC* cuts the outer scale.

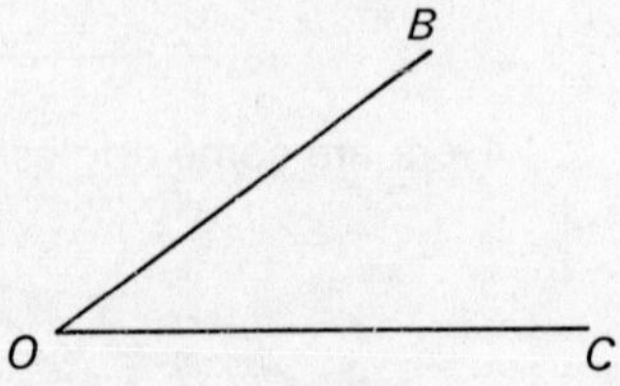

B

O C

Fig. 25

Here is another angle measured in two ways.

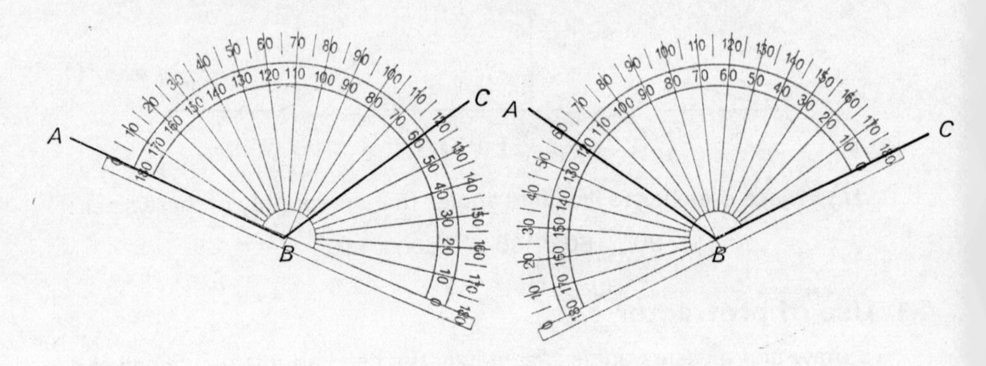

Fig. 26

4.1 Use of Protractor

(*a*) Figure 25 should be traced and the lines *OB* and *OC* extended, before any attempt at measurement is made.

The easiest way to find a reflex angle is to measure the acute or obtuse angle at the point and subtract your result from 360°.

Angles

Exercise D

1 $a = 40°$; $b = 58°$; $c = 122°$; $d = 20°$; $e = 82°$; $f = 38°$.

Since even the best of us occasionally read from the wrong scale, you should check your answer by seeing whether the angle is obtuse or acute, whether it 'looks right'.

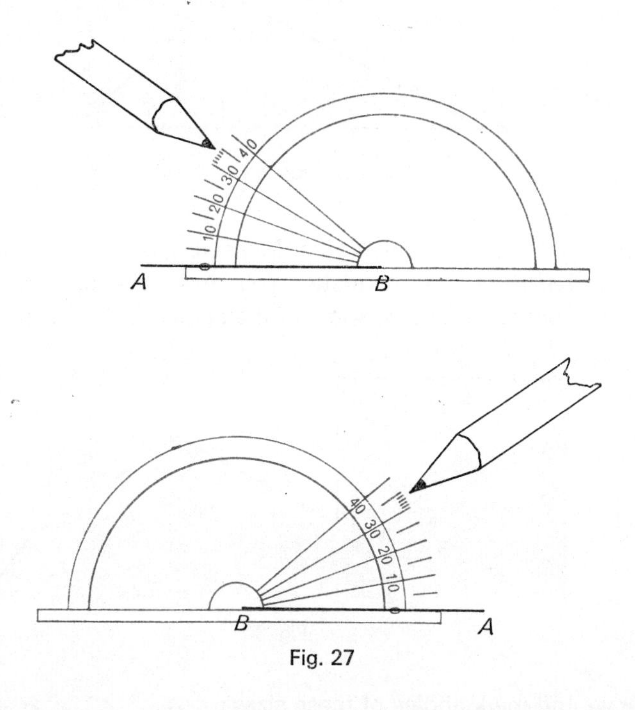

Fig. 27

(*b*) To draw an angle *ABC* of 34° mark the vertex *B*, and draw a line from it, ending at *A*. Place the centre of the protractor on *B*, and a zero line along *BA*. On the scale with this zero line find 34, and make a dot. Remove the protractor, and join the dot to *B*.

Make sure the angle you have drawn looks about the right size.

(*c*) How would you measure a reflex angle?

Exercise D

1 Measure the marked angles of this figure with your protractor.

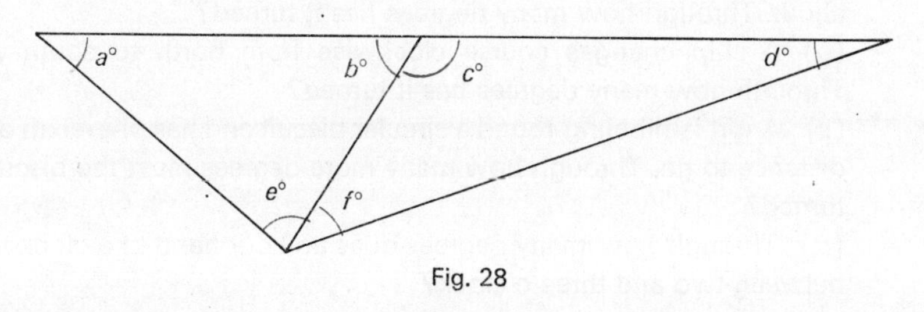

Fig. 28

2 Measure the following angles of the given figure: $\angle BAD$; $\angle BDC$; $\angle ABC$.

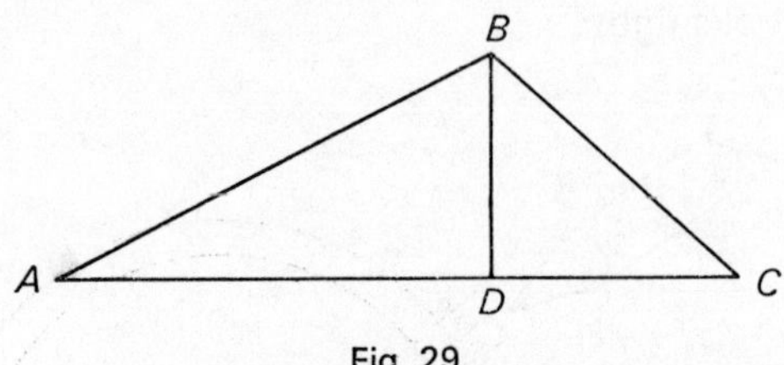

Fig. 29

3 Trace the outline of this figure. Measure the angles which are in the shaded part of the silhouette. You may have to extend the sides before you can measure some of the angles with your protractor.

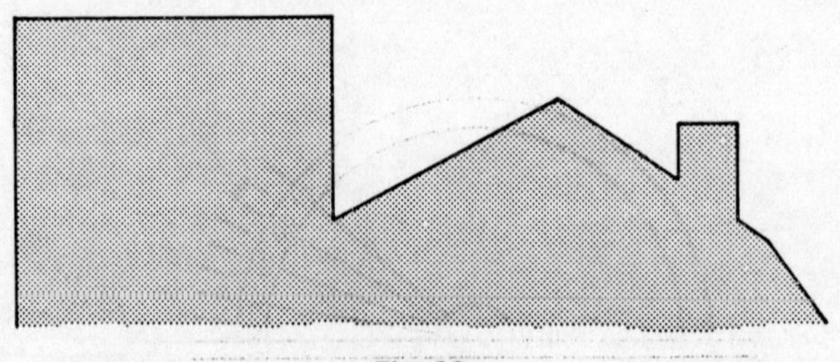

Fig. 30

4 Draw and mark angles of these sizes:

(*a*) 25°; (*b*) 44°; (*c*) 90°;
(*d*) 61°; (*e*) 120°; (*f*) 167°.

5 Draw two straight lines at the following angles as accurately as you can without using a protractor. (In time, you should be able to judge angles to within 20° by eye.) Check the angles with a protractor.

(*a*) 90°; (*b*) 60°; (*c*) 45°; (*d*) 30°; (*e*) 6°;
(*f*) 70°; (*g*) 120°; (*h*) 150°; (*i*) 100°; (*j*) 135°.

6 (*a*) A car has turned through a quarter-turn on its way round a roundabout. Through how many degrees has it turned?

(*b*) A ship changes course clockwise from north to south-west. Through how many degrees has it turned?

(*c*) A girl is nibbling round a circular biscuit and has one-sixth of the distance to go. Through how many more degrees must the biscuit be turned?

(*d*) Through how many degrees does an hour hand of a clock move between two and three o'clock?

2 $\angle BAD = 27°$; $\angle BDC = 90°$; $\angle ABC = 112°$.

3 90°, 90°, 297°, 120°, 303°, 90°, 90°, 237°, 160°.

6 (*a*) 90°.
(*b*) 225°.
(*c*) 60°.
(*d*) 30°.

7 (*a*) $a=90$; (*b*) $c=60$; (*c*) $b=60$; (*d*) $d=140$;
(*e*) $t=345$; (*f*) $a=140$; (*g*) $b=125$; (*h*) $s=91$;
(*i*) $v=179$; (*j*) $a=55$, $b=c=125$.

8 (*a*) Half-turn or 180°.
(*b*) They form a straight line.
(*c*) Yes.
(*d*) 10°, 170°; 55°, 125°; 75°, 105°; 100°, 80°; 145°, 35°.
(*e*) 10°, 55°.

7 In the following diagrams, right-angles and some other angles are marked. Calculate (do not use your protractor) the size of the angles marked with small letters.

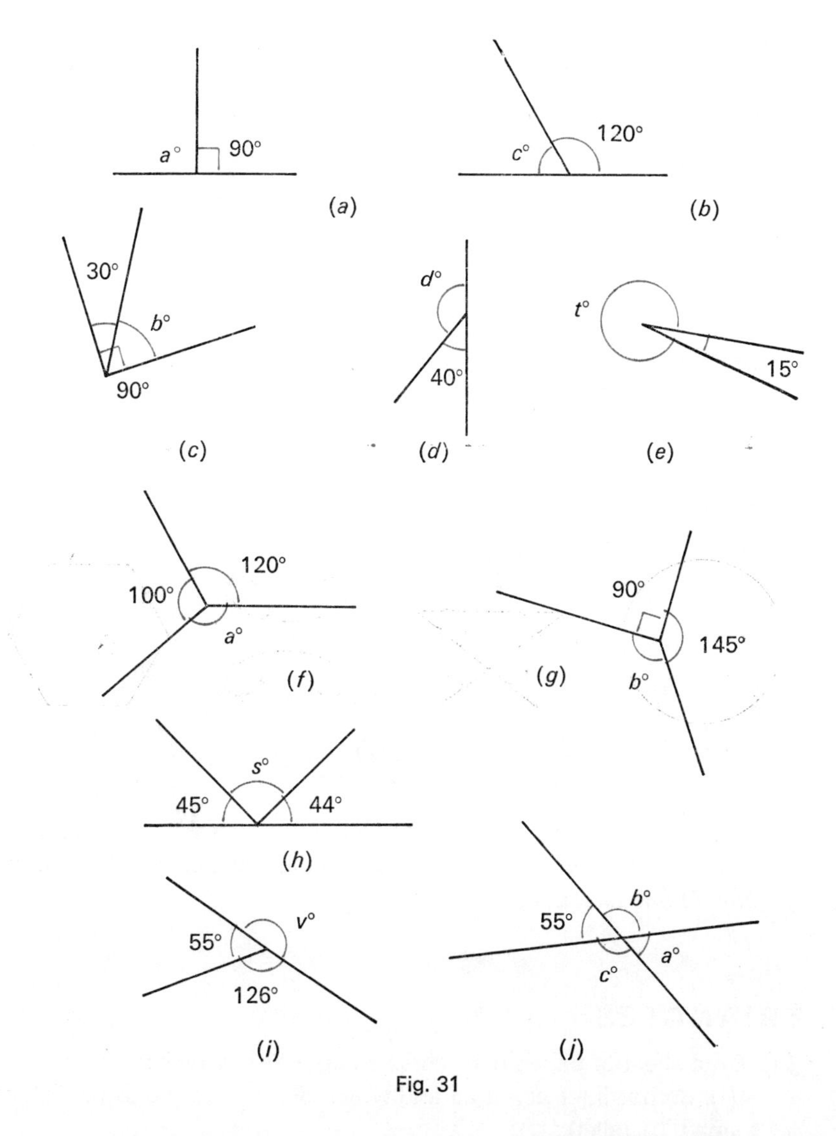

Fig. 31

8 (*a*) What is the sum of two angles, $\angle AOB$ and $\angle BOC$ at O, if AC is a straight line? (Such angles are called '*supplementary*' angles.)
(*b*) If two angles that are supplementary have a vertex and one line in common, what can you say about the other two lines?
(*c*) Two angles, $\angle AOB$ and $\angle BOC$, together form a right-angle. (They are said to be '*complementary*'.) Is OA perpendicular to OC?

(*d*) Here are two sets of angles:

$$A = \{10°, 55°, 75°, 100°, 145°\};$$

$$B = \{35°, 80°, 105°, 125°, 170°\}.$$

For each angle in set A, find its supplementary angle in set B.

(*e*) Which angles in set A have a complement in set B?

9 (*a*)

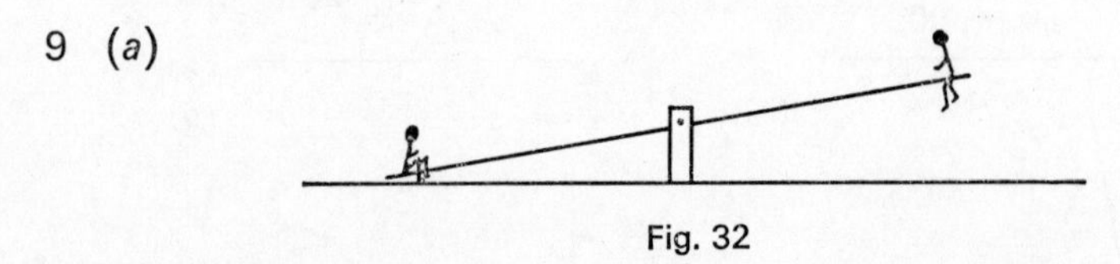

Fig. 32

Copy the picture of the seesaw in which the beam moves through 20°. Through what angle do the children turn? About what point are they rotating?

(*b*) *Experiment.* Draw the following curves on the floor so that they can be walked round.

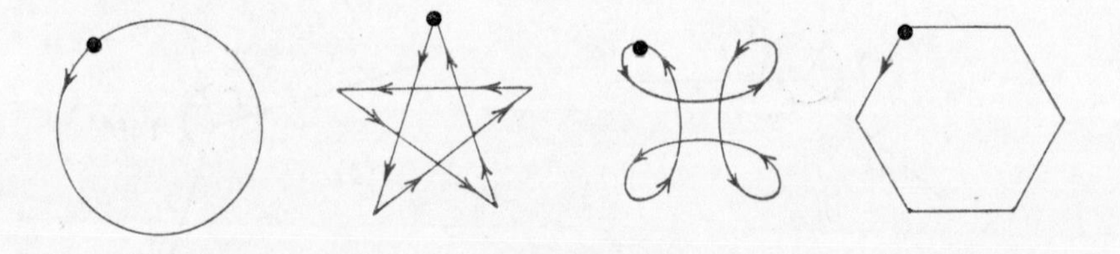

Fig. 33

Through how many whole turns have you rotated if you walk round each of these curves? Measure the amount which you have turned *only in one direction*.

5. TRIANGLES

(*a*) On a sheet of paper, draw three straight lines which:

(i) do not intersect at all and which are always the same distance apart (parallel);

(ii) intersect at only one point;

(iii) intersect at just two points;

(iv) intersect at three points.

Can you draw three straight lines which intersect at four points?

The figure formed when three straight lines intersect at three points is called a *triangle*. (Three angles.)

9 (*a*) Assuming that the children do not move relative to the beam, they also will move through 20°. The beam is rotating about the pivot at its centre and so are the children.

(*b*) 1, 2, 3, 1.

5. TRIANGLES

(*a*) No.

(*b*) The points must lie on a straight line.

(*c*) No. No. 3. 2.

(*d*) Yes. The triangular frame is the only one that is rigid.

(*e*) *ADC, ABD, BDC, BEC, DEC, AED, ABE.*

(*f*) (i) 180° is an ideal estimate. It is unlikely that this will be the result if the estimation comes from actual measurements.
(ii) The outside edges should form a straight line.

(*b*) Mark three points on a sheet of paper and join them with straight lines. What can you say about the position of the three points if the resulting figure is not a triangle?

(*c*) Draw a triangle with one obtuse angle. Can you draw a triangle with two obtuse angles? Can you draw a triangle with two right-angles?

What is the greatest number of acute angles that a triangle can have? What is the smallest number?

(*d*) Make a frame by threading cotton through four drinking straws. Can you alter the angles of the frame without bending or breaking the straws? Repeat the experiment, this time making a triangular frame. Can you alter the angles of the triangular frame?

The triangle is said to form a 'rigid' framework. How could you make the square rigid?

Engineers use the fact that a triangle forms a rigid framework in some of their constructions. Railway bridges, signal frames and some crane structures are examples.

(*e*)

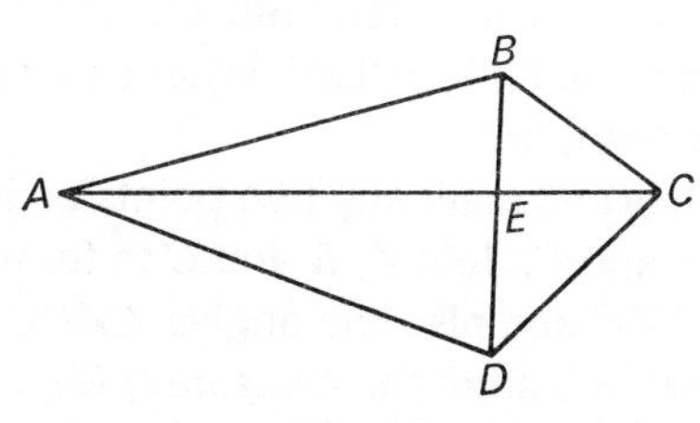

Fig. 34

A triangle is named using the three points that are its vertices. *ABC* is a triangle of this figure. Name the other seven.

(*f*) (i) Draw many different shaped triangles and measure their inside angles. What would be your estimate of the sum of the angles of a triangle? Would this be true of all triangles?

(ii) Draw a triangle and cut it out. Tear off the corners and arrange them as shown below. What do you notice about the sum of the angles at these corners? Does this agree with your estimate in (i)?

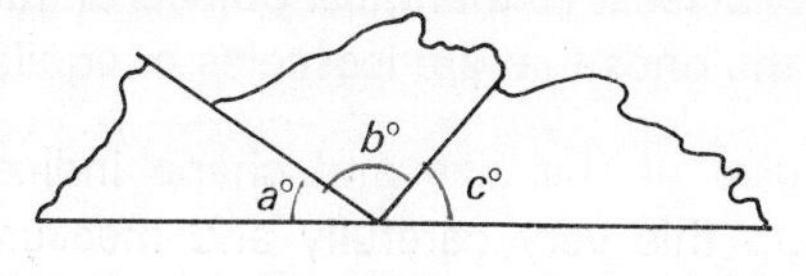

Fig. 35

(iii)

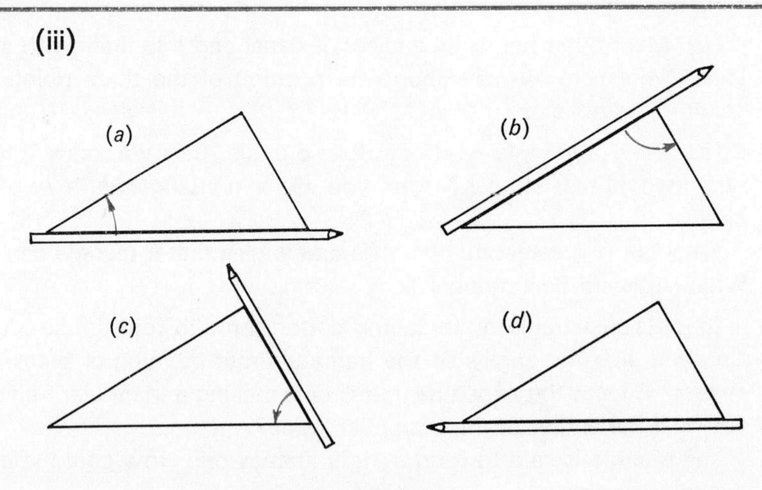

Fig. 36

Place a pencil along one side of a triangle and turn it about each vertex in order as shown in the diagram. What do you notice about the final position of the pencil compared with its first position? Through what angle has it been turned altogether? What can you say about the sum of the angles of the triangle?

(*g*) Draw a circle and mark any two points on it. Call the centre *A*, and the two points, *B* and *C*. Join *A*, *B* and *C* to form a triangle and measure the sides *AB* and *AC* and also the angles $\angle ABC$ and $\angle ACB$. (Is there a case where there are no angles to measure?) Repeat the construction with different sizes of circle and with different positions for *B* and *C*. The sides *AB* and *AC* will always be the same length. What can you say about the angles opposite these equal sides?

A triangle with two equal sides is called *isosceles*.

Would you expect a triangle with two equal angles also to have equal sides? Construct some triangles with equal angles and see.

The length *BC* could be made to equal the length of the other two sides. This special isosceles triangle is called an *equilateral* triangle. What is the size of each angle of an equilateral triangle?

Exercise E

1 Make a list of at least 10 triangular objects or frameworks that you have seen. Note the ones that are isosceles or equilateral.

2 Draw triangles of the size and shape indicated on the following diagrams. Do this very carefully and measure the other sides and angles in each case.

(iii) The pencil will have been rotated through a half-turn and this gives the sum of the angles of the triangle.

(*g*) There are no angles to measure if *A*, *B* and *C* lie on a straight line. The radii are always equal and the angles opposite these radii are equal. You would expect that if you constructed a triangle with two equal angles, the sides opposite to them would be equal. The angles of an equilateral triangle are each 60°.

Exercise E

1 Triangle, the musical instrument (equilateral); roof supports and roofs of houses; scaffolding and building structures; postage stamps; folded table napkins; signal flags (isosceles); road signs (mostly equilateral); sails of yachts; sandwiches; ends of ramps; five-bar gates. Also, the 'Union Jack' contains many triangles.

2 (*a*) 2·33 cm; 41·8°; 87·2°.
(*b*) 75°; 2·42 cm; 1·48 cm.
(*c*) 30°; 3·28 cm; 1·69 cm.
(*d*) 42°; 103°; 5·1 cm.

3 (*a*) $a=70$; (*b*) $b=34$; (*c*) $c=60$;
(*d*) $d=45$; (*e*) $e=90$; (*f*) $f=30$.

4 In each case the marked angle equals the sum of the interior opposite angles.
(*a*) 90; (*b*) 112; (*c*) 5; (*d*) 156; (*e*) 78.

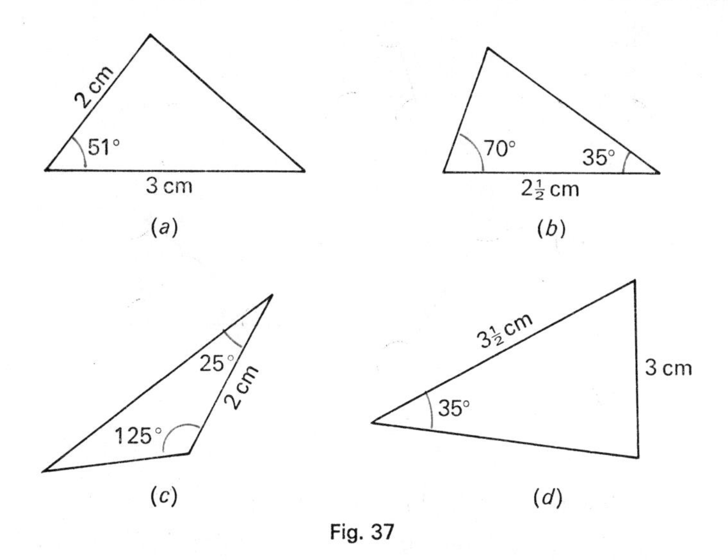

Fig. 37

3 Assuming that the three angles of a triangle add up to 180°, work out the size of the third angle in the following triangles:

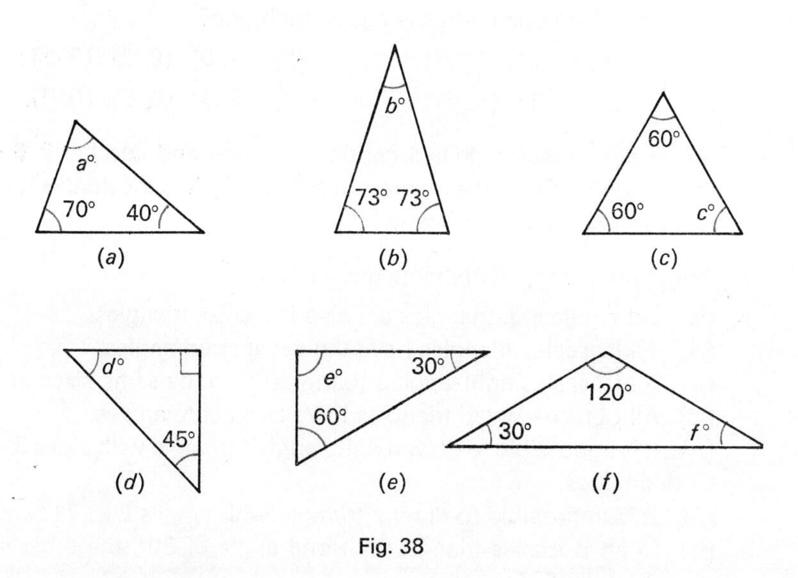

Fig. 38

4 Work out the size of the angle marked with a letter in each of the cases in Figure 39. Try to find a direct connection between the lettered angle and the two given angles in each case.

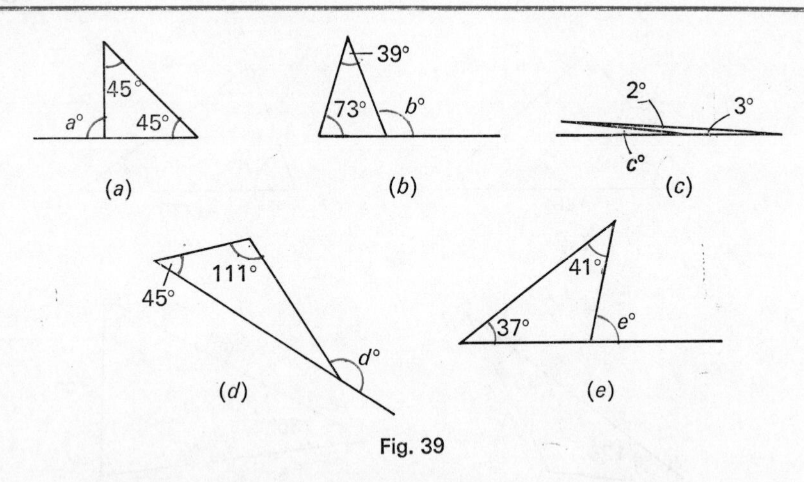

Fig. 39

5 If possible, draw each of the following:

(*a*) a triangle with one obtuse inside angle;
(*b*) a triangle with two obtuse inside angles;
(*c*) a triangle with two right-angles;
(*d*) a triangle with three acute angles.

6 Draw the triangles whose vertices are given by the following co-ordinates. What can you say about each one?

(*a*) (0, 0), (2, 0), (0, 2); (*b*) (3, 0), (6, 2), (5, 3);
(*c*) (0, 3), (2, 1), (5, 4); (*d*) (2, 4), (5, 5), (0, 5).

7 If one side of a triangle is 4 centimetres long and another is 6 centimetres long, what is the greatest possible length of the third side? What is its smallest possible length?

8 Are the following statements true or false?

(*a*) All equilateral triangles are also isosceles triangles.
(*b*) All isosceles triangles have the same sized angles.
(*c*) All isosceles right-angled triangles have the same sized angles.
(*d*) All obtuse-angled triangles have two acute angles.
(*e*) It is impossible to draw a right-angled triangle with sides 3, 4 and 5 centimetres.
(*f*) It is impossible to draw a triangle with angles 24°, 71° and 63°.
(*g*) In an isosceles triangle with one angle of 20°, the other angles must both be 80°.
(*h*) All three angles of an equilateral triangle are the same size.
(*i*) Three different triangles can be drawn with one angle of 20°, another of 30° and one side 8 centimetres long.

5 (*b*) and (*c*) are not possible.

6 (*a*) An isosceles, right-angled triangle.
(*b*) An isosceles triangle.
(*c*) A right-angled triangle.
(*d*) A triangle without regularity (scalene).

7 The greatest possible length is just less than 10 cm and the least is just more than 2 cm.

8 (*a*) True; (*b*) false; (*c*) true;
(*d*) true; (*e*) false; (*f*) true;
(*g*) false; (*h*) true; (*i*) true.

4. Number bases

The purpose of this chapter is to revise the basic concepts and operations of arithmetic and particularly to draw attention to the importance of position in our method of writing numbers. It is not considered that the ability to count or calculate in bases other than ten is important in itself. In particular, there is no point in learning tables in other bases; when necessary pupils should have access to previously prepared tables. There will be a second chapter on number bases in Book B, which deals with binary and duodecimal systems and their applications.

1. COUNTING NUMBERS

This first section is not to be laboured, but it is sometimes found useful to remind pupils that counting is the process of establishing a one-to-one correspondence between the objects which are being counted and an arbitrary set of symbols. In this section, pupils should be encouraged to invent their own sets of numerals and it is hoped that a discussion will arise between those who use number place systems and those who do not, as to the relative merits of the two.

The number line has been introduced, not because it is necessary in the process of counting, but because we shall refer to it throughout the course. The points representing counting numbers are then associated with other points to form a continuous line, though we shall not make any effort to explain that rational and irrational numbers are required to do this. Negative numbers will be introduced in a chapter in Book B.

4. Number bases

1. COUNTING NUMBERS

1.1 Counting

How many stars are there?

To answer this question, we have to count. When we count the number of objects in a collection, we mentally tick off each object against this list:

1, 2, 3, 4, 5, 6, 7, 8, 9, ...

We can think of this list as showing a series of equal steps along a line.

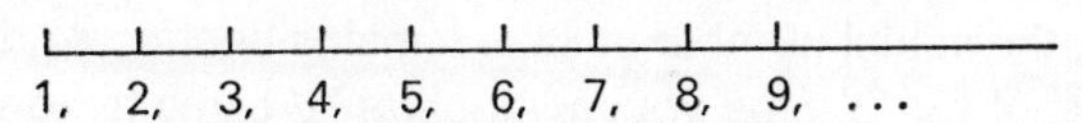

We can place the stars along the line like this

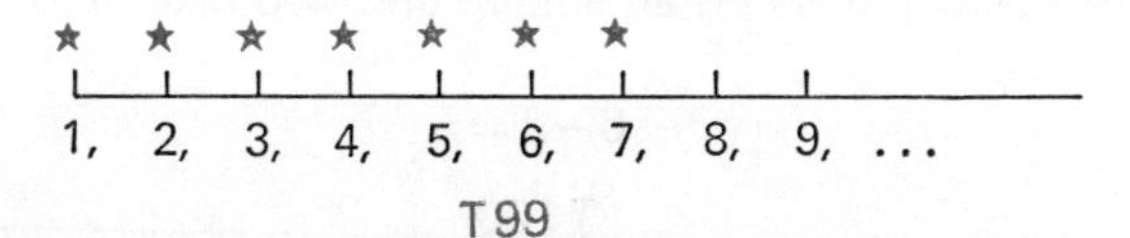

T99

Because the last star is beside the '7', we say that there are 'seven' stars in the group. There will be times when we go to count a set of objects and find that there are none. So we shall add '0' to our set of numbers on the line which will now look like this:

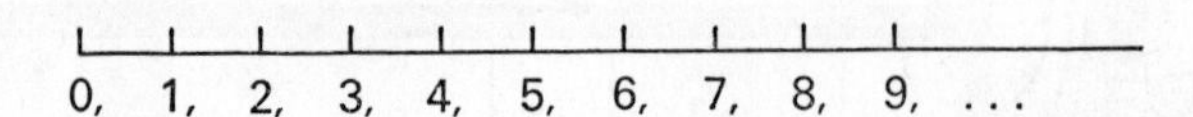

Do the symbols that we use make any difference to the position of the last star along the number line? Is this position altered if we write different numerals along the line? Here are some examples of other systems of numerals:

	1	2	3	4	5	10	12
Babylonian	𒁹	𒈫	𒐈	𒐉	𒐊	𒌋	𒌋𒈫
Egyptian hieratic	I	II	III	IIII	[illegible]	[illegible]	[illegible]II
Early Roman	I	II	III	IIII	V or Λ	X	XII

Invent a series of number symbols for yourself and give them names. Try counting collections of objects so that you get used to your own system. Now use your system to add and subtract.

1.2 Addition and subtraction

From the picture, you can see that whatever symbol you use in the addition, the actual number of stars remains unaffected. Did you find that your symbol for '3' plus your symbol for '2' came to your symbol for '5'? See if your system works for other additions.

If there are six planes on an airstrip and two take off, there will be four left. We write

$$6-2=4.$$

1.2 and 1.3 Addition, subtraction and multiplication

Pupils should be encouraged to carry out these operations in their own number system especially with numbers large enough to use the processes of carrying or borrowing, where these are appropriate.

Reference to repeated subtraction will again be made in the chapter on division, where the problems of expressing the remainders will be discussed.

On the number line this would be shown as

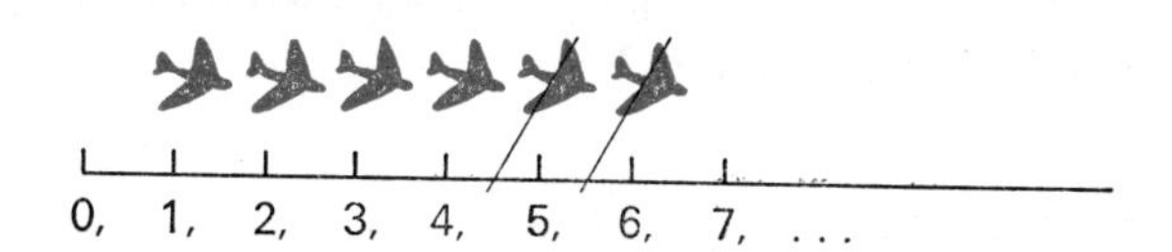

Can you subtract using your system of symbols?

1.3 Multiplication

Multiplication is a short way of counting objects when we have several groups of them that are all the same size.

This flight of geese is in three groups, each of five birds. If we mark these off in groups along the number line, we would have

5 5 5

0, 1, 2, 3, 4, 5, 6, 7, 8, 9, 10, 11, 12, 13, 14, 15, 16, . . .

We have repeatedly added fives together. In number symbols this repeated addition is written

$$5+5+5 = 15.$$

However, it is quicker to learn and remember that three lots of five come to fifteen ($3 \times 5 = 15$), and so we have our multiplication tables.

Just as multiplication is repeated addition, so division may be thought of as repeated subtraction. If we start with twelve dogs and take them off in pairs for a walk, the number of walks we would have to take would be given by twelve divided by two. On the number line, this can be shown as

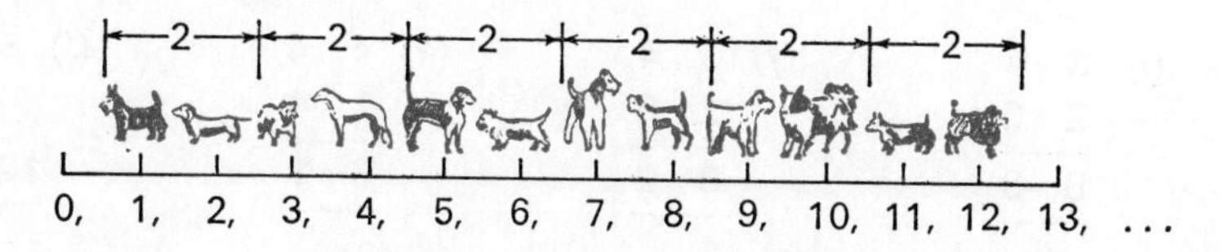

Can you use *your* system to carry out multiplication and division? Do the symbols that you have used affect the actual number of stars, planes, geese or dogs?

What is it about the standard number system that makes it so generally used? Why is it that the next numbers after nine have two digits? Did they have two digits in your system?

2. WORKING IN COLUMNS

The text and exercises in Section 2 involve the use of imperial units.

2.1 Addition

Which of these is correct? Add them up and see.

$$(a)\ \begin{array}{r} 4\ 7 \\ +\ 3\ 9 \\ \hline 8\ 4 \\ \hline \end{array} \qquad (b)\ \begin{array}{r} 4\ 7 \\ +\ 3\ 9 \\ \hline 8\ 0 \\ \hline \end{array} \qquad (c)\ \begin{array}{r} 4\ 7 \\ +\ 3\ 9 \\ \hline 8\ 6 \\ \hline \end{array}$$

You have probably found that (*c*) is right, but in fact all three could be.

In (*a*) put feet and inches as column headings.

In (*b*) put pounds and ounces as column headings.

In (*c*) put tens and units as column headings.

Now add up and check again.

Can you think of any other column headings which would make (*a*) correct?

We usually think that numbers in columns represent hundreds, tens, and units, but this need not be the case.

Exercise A

1 Find column headings which will make these additions correct:

$$(a)\ \begin{array}{r} 2\ 8 \\ 1\ 9 \\ \hline 4\ 5 \\ \hline \end{array} \qquad (b)\ \begin{array}{r} 2\ 8 \\ 1\ 9 \\ \hline 4\ 1 \\ \hline \end{array} \qquad (c)\ \begin{array}{r} 2\ 8 \\ 1\ 9 \\ \hline 4\ 7 \\ \hline \end{array} \qquad (d)\ \begin{array}{r} 2\ 9 \\ 3\ 7 \\ \hline 6\ 0 \\ \hline \end{array}$$

$$(e)\ \begin{array}{r} 2\ 9 \\ 3\ 7 \\ \hline 6\ 2 \\ \hline \end{array} \qquad (f)\ \begin{array}{r} 4\ 9 \\ 3\ 5 \\ \hline 8\ 2 \\ \hline \end{array} \qquad (g)\ \begin{array}{r} 2\ 2 \\ 1\ 2 \\ \hline 4\ 1 \\ \hline \end{array} \qquad (h)\ \begin{array}{r} 2\ 3 \\ 5\ 8 \\ \hline 8\ 1 \\ \hline \end{array}$$

$$(i)\ \begin{array}{r} 3\ 4 \\ 2\ 6 \\ \hline 6\ 3 \\ \hline \end{array} \qquad (j)\ \begin{array}{r} 3\ 4 \\ 2\ 6 \\ \hline 6\ 2 \\ \hline \end{array} \qquad (k)\ \begin{array}{r} 4\ 4 \\ 1\ 4 \\ \hline 6\ 1 \\ \hline \end{array} \qquad (l)\ \begin{array}{r} 8\ 6 \\ 4\ 2 \\ \hline 12\ 8 \\ \hline \end{array}$$

$$(m)\ \begin{array}{r} 3\ 8 \\ 2\ 8 \\ \hline 6\ 0 \\ \hline \end{array} \qquad (n)\ \begin{array}{r} 2\ 8 \\ 7 \\ \hline 3\ 1 \\ \hline \end{array} \qquad (o)\ \begin{array}{r} 5\ 2 \\ 3\ 2 \\ \hline 9\ 1 \\ \hline \end{array}$$

2. WORKING IN COLUMNS

2.1 Addition

Although metric units will be used throughout this course, it is considered that, for some years, pupils will be sufficiently well acquainted with imperial units to profit from the approach. The section starts with an exercise intended to jolt the pupil out of always thinking in hundreds, tens and units by appealing to quantities such as feet and inches, pounds and ounces.

These are not strictly examples of other *number* bases; for example, pints and quarts do not really represent a base two system since you can write 3 qt, 1 pt. The digit 3 cannot appear in base two.

Exercise A

1 (*a*) Ft and in., years and months;
(*b*) lb and oz; (*c*) tens and units; (*d*) lb and oz;
(*e*) st and lb; (*f*) ft and in., years and months;
(*g*) yd and ft; (*h*) tens and units; (*i*) weeks and days;
(*j*) gal and pt, miles and furlongs; (*k*) weeks and days;
(*l*) tens and units; (*m*) lb and oz; (*n*) st and lb;
(*o*) yd and ft.

2.2 Subtraction

The difficulty of subtraction in this context makes the recognition of an inverse process a very real advantage and this should be stressed as a foundation for later work.

2.2 Subtraction

It is not so easy to discover the column headings if they are missing from a subtraction.

(*a*) Try this:

$$\begin{array}{r} 6\ 2 \\ -\ 1\ 9 \\ \hline 4\ 5 \\ \hline \end{array}$$

You would start by saying '2, take away 9'. This cannot be done, but the next idea should help to make things easier.

Take an easy example first:

$$\begin{array}{r} 7 \\ -\ 5 \\ \hline 2 \\ \hline \end{array}$$

You can always check a subtraction by adding together the answer and the number you took away; this should give you the number you started with.

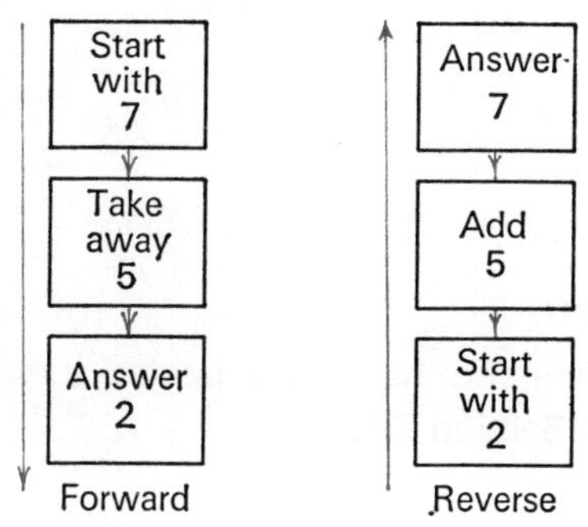

See how this works with:

$$\begin{array}{r} \text{t}\ \text{u} \\ 4\ 7 \\ -\ 2\ 9 \\ \hline 1\ 8 \\ \hline \end{array} \quad \text{which becomes} \quad \begin{array}{r} \text{t}\ \text{u} \\ 1\ 8 \\ +\ 2\ 9 \\ \hline 4\ 7 \\ \hline \end{array}$$

So you see, we can always check a subtraction question by turning it into an addition question.

(*b*) Find column headings for this subtraction question by reversing and then checking the addition:

$$\begin{array}{r} 4\ 2 \\ -\ 2\ 9 \\ \hline 1\ 7 \\ \hline \end{array} \quad \text{can be written as} \quad \begin{array}{r} 1\ 7 \\ +\ 2\ 9 \\ \hline 4\ 2 \\ \hline \end{array}$$

which should show you that you are dealing with stones and pounds, so put in the column headings and check,

$$\begin{array}{rcc} & \text{st} & \text{lb} \\ & 4 & 2 \\ - & 2 & 9 \\ \hline & 1 & 7 \\ \hline \end{array}$$

Exercise B

1 Find the missing numbers:

(*a*) $\begin{array}{rcc} & t & u \\ & ? & ? \\ - & 1 & 3 \\ \hline & 2 & 5 \\ \hline \end{array}$ (*b*) $\begin{array}{rcc} & t & u \\ & ? & ? \\ - & 3 & 6 \\ \hline & 4 & 1 \\ \hline \end{array}$ (*c*) $\begin{array}{rcc} & t & u \\ & ? & ? \\ - & 4 & 7 \\ \hline & 2 & 8 \\ \hline \end{array}$ (*d*) $\begin{array}{rcc} & t & u \\ & ? & ? \\ - & 5 & 6 \\ \hline & 3 & 6 \\ \hline \end{array}$

2 Find the column headings:

(*a*) $\begin{array}{rcc} & 4 & 2 \\ - & 2 & 9 \\ \hline & 1 & 5 \\ \hline \end{array}$ (*b*) $\begin{array}{rcc} & 5 & 4 \\ - & 2 & 7 \\ \hline & 2 & 5 \\ \hline \end{array}$ (*c*) $\begin{array}{rcc} & 2 & 0 \\ - & 1 & 1 \\ \hline & & 1 \\ \hline \end{array}$ (*d*) $\begin{array}{rcc} & 2 & 2 \\ - & 1 & 6 \\ \hline & & 3 \\ \hline \end{array}$

2.3 Multiplication

Multiplication is a shorthand for repeated addition and the same ideas can be used as in Section 2.1.

Exercise C

1 Find the column headings:

(*a*) $\begin{array}{rcc} & 2 & 7 \\ \times & & 3 \\ \hline & 7 & 5 \\ \hline \end{array}$ (*b*) $\begin{array}{rcc} & 2 & 7 \\ \times & & 3 \\ \hline & 7 & 9 \\ \hline \end{array}$ (*c*) $\begin{array}{rcc} & 2 & 7 \\ \times & & 3 \\ \hline & 7 & 1 \\ \hline \end{array}$ (*d*) $\begin{array}{rcc} & 2 & 7 \\ \times & & 3 \\ \hline & 7 & 7 \\ \hline \end{array}$

(*e*) $\begin{array}{rcc} & 2 & 7 \\ \times & & 3 \\ \hline & 8 & 1 \\ \hline \end{array}$ (*f*) $\begin{array}{rcc} & 2 & 7 \\ \times & & 3 \\ \hline & 9 & 0 \\ \hline \end{array}$ (*g*) $\begin{array}{rcc} & 1 & 2 \\ \times & & 2 \\ \hline & 3 & 0 \\ \hline \end{array}$ (*h*) $\begin{array}{rcc} & 2 & 6 \\ \times & & 4 \\ \hline & 9 & 8 \\ \hline \end{array}$

(*i*) $\begin{array}{rc} & 1 \\ \times & 5 \\ \hline & 21 \\ \hline \end{array}$ (*j*) $\begin{array}{rc} & 3 \\ \times & 5 \\ \hline & 11 \\ \hline \end{array}$

Exercise B

1 (*a*) 38; (*b*) 77; (*c*) 75; (*d*) 92.

2 (*a*) Ft and in., years and months;
(*b*) gal and pt, miles and furlongs;
(*c*) qt and pt; (*d*) weeks and days.

2.3 Multiplication

Addition is easy, subtraction is not, and inverse processes are used. Next we come to multiplication and find that this also is easy; the question as to why this is so should be raised here, and the idea of multiplication as repeated addition discussed.

Exercise C

1 (*a*) Lb and oz; (*b*) ft and in., years and months;
(*c*) ton and cwt; £ and sh; (*d*) st and lb;
(*e*) tens and units; (*f*) weeks and days;
(*g*) cwt and qr; (*h*) lb and oz;
(*i*) qt and pt; (*j*) st and lb.

2.4 Division

After the discussion suggested for Section 2.3 above, it is hoped that the pupils will be able to explain why division is so hard and needs inverse treatment.

Exercise D

1 (*a*) 96; (*b*) 84; (*c*) 95.

2 (*a*) Ton and cwt; £ and sh; (*b*) gal and pt; miles and furlongs;
(*c*) st and lb; (*d*) yd and ft;
(*e*) weeks and days.

3. GROUPING

3.1 Base ten

Exercise E should not be 'skipped'. It will take a little time, and apart from a welcome break from figures, it will give the opportunity to think of what is involved here. Moreover, the style and ingenuity of the answers can be very interesting and will often provoke discussion.

2.4 Division

Division is more difficult but, just as with subtraction, we can find a way round it. Take an easy example first.

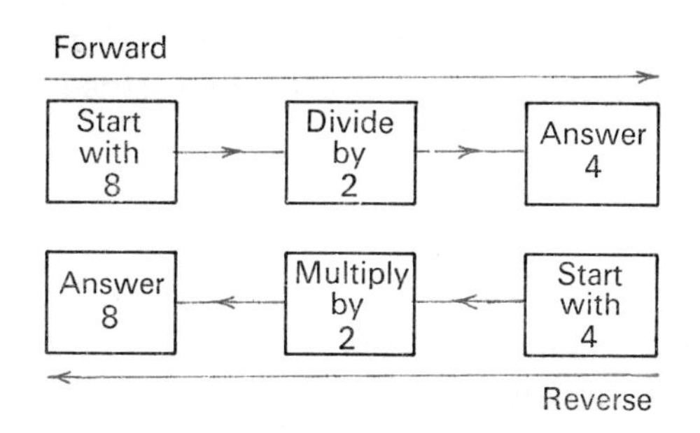

We can use this idea to find the column headings for:

$$\begin{array}{r} 1\ \ 3 \\ 6\overline{)7\ \ 6} \end{array}, \quad \text{by changing the question into} \quad \begin{array}{r} 1\ \ 3 \\ \times\quad 6 \\ \hline 7\ \ 6 \\ \hline \end{array}$$

and you should now see that it could be feet and inches.

Exercise D

1 Find the missing numbers:

(*a*) $\begin{array}{r} \text{t}\ \ \text{u} \\ 3\ \ 2 \\ 3\overline{)?\ \ ?} \end{array}$ (*b*) $\begin{array}{r} \text{t}\ \ \text{u} \\ 1\ \ 4 \\ 6\overline{)?\ \ ?} \end{array}$ (*c*) $\begin{array}{r} \text{t}\ \ \text{u} \\ 1\ \ 9 \\ 5\overline{)?\ \ ?} \end{array}$

2 Find the column headings:

(*a*) $\begin{array}{r} 1\ \ 4 \\ 6\overline{)7\ \ 4} \end{array}$ (*b*) $\begin{array}{r} 1\ \ 2 \\ 6\overline{)7\ \ 4} \end{array}$ (*c*) $\begin{array}{r} 1\ \ 3 \\ 6\overline{)7\ \ 4} \end{array}$

(*d*) $\begin{array}{r} 1\ \ 1 \\ 4\overline{)5\ \ 1} \end{array}$ (*e*) $\begin{array}{r} 1\ \ 3 \\ 5\overline{)7\ \ 1} \end{array}$

3. GROUPING

3.1 Base ten

All the examples so far were meant to remind you that numbers in columns really need headings. It is the custom to leave out the headings when we are working in tens.

Why is it that we normally count things in groups of ten? Nobody knows for certain, but it seems very likely that it is because we have ten fingers and people use their fingers to count with.

Early man would come out of his cave in the morning and tally his sheep against his fingers to see if the wolves had taken any sheep overnight. When all ten fingers had been used up, he 'carried' this fact either in his head or by making a mark and then he started to use his fingers all over again.

Exercise E

1 Make a drawing of a cave man counting his sheep,

(*a*) if he had only four sheep;

(*b*) if he had twelve sheep.

3.2 Other bases

Things might have been very different if man had only one arm with five fingers on the hand. Then he would have to record every group of five instead of every group of ten.

This number of sheep would be thought of as or as

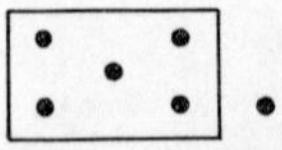

The 'one' that was carried stood for a group of five units. Writing this in number symbols we could put 12.

But 12 might be confused with twelve, so we must read it as 'one two' and write it either as

Fives	Units
1	2

or, more neatly, as 12_{five}.

In future work we shall use this method and call it *base five*.

Exercise F

1 Copy these patterns of dots. Count them in base ten, then group into fives and write the answers in number symbols to base five.

(*a*)  (*b*)

(*c*)  (*d*)

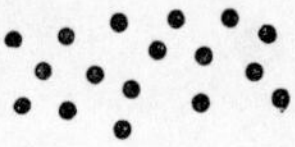

(*e*) 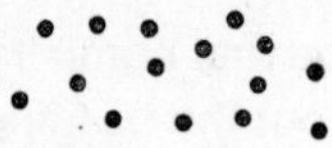 (*f*)

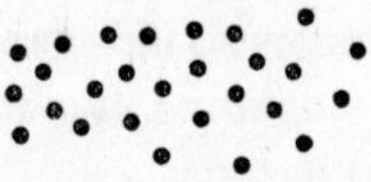

3.2 Other bases

Care must be exercised in the use of symbols and words for numbers in other bases; throughout the course, *words* are used in base ten in the normal sense, for example:

The English word for the NUMBER of dots is THIRTEEN.

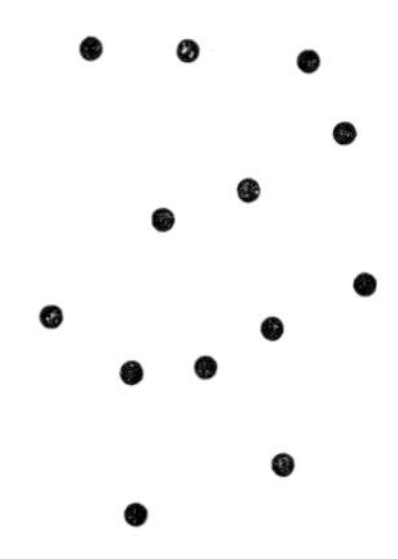

The number and the name are quite invariant for this particular set of dots.

However, if we wish to represent this number in symbols, or numerals, we can write:

13_{ten}, 15_{eight}, 21_{six}, 1101_{two} and so on.

If we wish to speak of these in words, we shall say, 'One, three, base ten', 'One, five, base eight', 'Two, one, base six', 'One, one, nought, one, base two'.

Naturally, in cases where the whole exercise is in one particular base, initial instructions will be given to avoid the use of a base suffix every time. In most work, the ordinary base ten system and notation will be used without any special comment.

Attempts have been made to invent words to describe numbers to bases other than ten, such as 'twocty-one' for 21_{eight}. Although these can sometimes be amusing, there appears to be little other value in this.

See: *An Introduction to Number Scales and Computers,* by F. J. Budden; also, *The Language of Mathematics,* by F. Land.

Exercise F

Tracing patterns of dots can easily damage a book and it is suggested that these patterns might be duplicated by the teacher and issued for use.

1 (*a*) 13_{five}; (*b*) 22_{five}; (*c*) 4_{five}; (*d*) 32_{five}; (*e*) 30_{five}; (*f*) 102_{five}.

2 (*a*) nine; (*b*) ten; (*c*) eighteen;
(*d*) twenty-one; (*e*) twenty-nine.

4. THE SPIKE ABACUS

It is very important that a set of abaci should actually be used. Balsa is suggested so that the spikes can be driven in without the need for drilling. No doubt schools which have workshops will use other and more permanent woods and this will also be cheaper. For schools without facilities it is suggested that a local model shop be asked to supply a set of 15 cm × 4 cm × 2·5 cm blocks cut to size; these should be about 3p each. Balsa has the advantage that the spikes can be easily withdrawn and replaced with ones of different lengths and still remain tight in the holes.

Cutting cotton reels in half can be a frustrating task and it is suggested that a vice is used or something like a screwdriver is jammed in the hole to stop it rolling while the cutting takes place. Large drilled beads are better if funds are available.

The teacher will find a demonstration model quite essential. The Educational Supply Association market a Multi-Base Abacus (9898/869) which can be adjusted to suit many bases.

If an overhead projector is available, the abacus can be drawn on acetate sheet and coins or counters used directly.

2 Show these base five numbers as groups of dots.

(*a*) 14_{five}; (*b*) 20_{five}; (*c*) 33_{five};

(*d*) 41_{five}; (*e*) 104_{five}.

4. THE SPIKE ABACUS

All the ideas of this section need to be illustrated by using a spike abacus. This is so important that it is worth making one, even roughly, if there are none ready to be used.

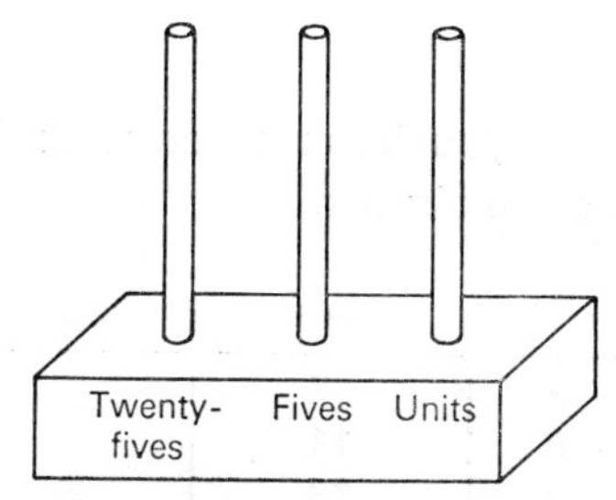

4.1 Using the spike abacus

We shall use the spike abacus to count in fives. Mark the spikes 'units', 'fives', 'twenty-fives'.

As you count, go on adding rings to the units spike until you get:

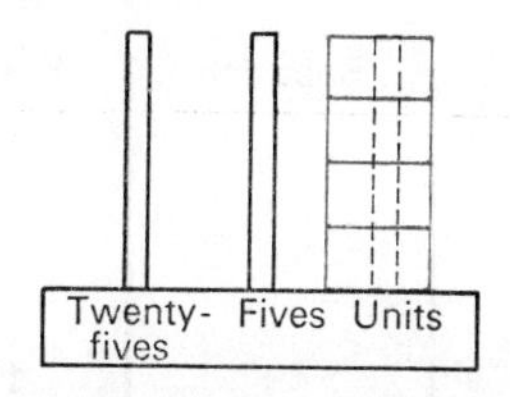

If you try to add another, there is no spike left for it to go on, so 'carry' one ring on to the fives spike and clear the units:

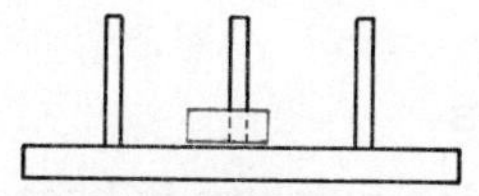

Start all over again on the units spike and continue until you reach:

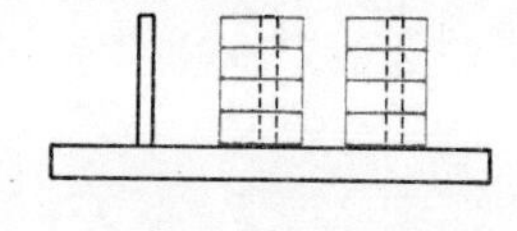

You should now have counted twenty-four. If you now try to add another to the units spike you have to carry one to the fives spike, but this is full up also, so you will carry again and end up with:

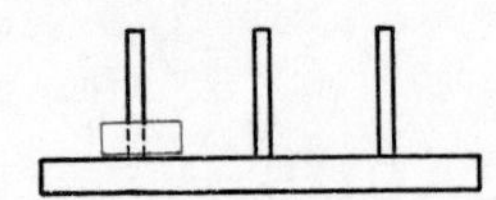

What is the greatest number you can represent on this spike abacus?

Exercise G

1 Copy this table into your book and complete it. Do this by counting the rings on to the spike abacus.

	Base ten	Base five
a	7	
b	14	
c	26	
d	38	
e	57	
f	100	

2 Load the abacus with rings to show these numbers in base five and count them in base ten as they are removed.

	Base ten	Base five
a		11
b		3
c		40
d		34
e		111
f		322

Copy this table into your book and complete it.

4.2 Operation tables

(*a*) Add the following numbers in base five and give their sum in the same base. If necessary, use your abacus.

$\begin{array}{r}1\\+1\\\hline\\\hline\end{array}$ $\begin{array}{r}1\\+2\\\hline\\\hline\end{array}$ $\begin{array}{r}1\\+3\\\hline\\\hline\end{array}$ $\begin{array}{r}1\\+4\\\hline\\\hline\end{array}$ $\begin{array}{r}2\\+2\\\hline\\\hline\end{array}$ $\begin{array}{r}2\\+3\\\hline\\\hline\end{array}$ $\begin{array}{r}2\\+4\\\hline\\\hline\end{array}$ $\begin{array}{r}3\\+3\\\hline\\\hline\end{array}$ $\begin{array}{r}3\\+4\\\hline\\\hline\end{array}$ $\begin{array}{r}4\\+4\\\hline\\\hline\end{array}$

Exercise G

1

	Base ten	Base five
a	7	12
b	14	24
c	26	101
d	38	123
e	57	212
f	100	400

2

	Base ten	Base five
a	6	11
b	3	3
c	20	40
d	19	34
e	31	111
f	87	322

4.2 Operation tables

With the addition table for base five there is an opportunity to consider the implication of the pattern in the array of numbers. The commutative law evidently holds, for the pattern is symmetrical about the leading diagonal.

Note also the inverse process again for subtraction.

Exercise H

All answers in base five.

(*a*) 22;	(*b*) 21;	(*c*) 31;	(*d*) 42;
(*e*) 34;	(*f*) 101;	(*g*) 100;	(*h*) 140;
(*i*) 240;	(*j*) 322;	(*k*) 422;	(*l*) 11;
(*m*) 3;	(*n*) 3;	(*o*) 14;	(*p*) 24.

(*b*) We can use these results to make out an 'operations' table for addition of numbers written in base five.

+	0	1	2	3	4
0	0	1	2	3	4
1	1	2	3	4	10
2	2	3	4	10	11
3	3	4	10	11	12
4	4	10	11	12	13

Copy the table into your book for reference. Can you see any patterns among the numbers of the table?

(*c*) We can use this table for subtraction in the following way:

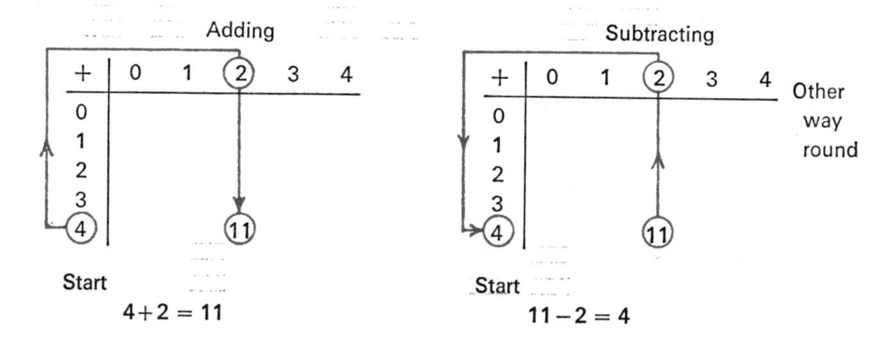

Check all the entries in the table with your spike abacus.

Exercise H

Perform the following additions and subtractions where all the numbers are in base five and give your answers in base five also. Use the operation table where necessary.

1 (*a*) 14 + 3 (*b*) 12 + 4 (*c*) 23 + 3 (*d*) 4 + 33

(*e*) 32 + 2 (*f*) 31 + 20 (*g*) 44 + 1 (*h*) 124 + 11

(*i*) 213 + 22 (*j*) 231 + 41 (*k*) 343 + 24 (*l*) 23 − 12 (*m*) $\begin{array}{r} 10 \\ -\ \ 2 \\ \hline \\ \hline \end{array}$ (*n*) $\begin{array}{r} 14 \\ -\ 11 \\ \hline \\ \hline \end{array}$ (*o*) $\begin{array}{r} 33 \\ -\ 14 \\ \hline \\ \hline \end{array}$ (*p*) $\begin{array}{r} 42 \\ -\ 13 \\ \hline \\ \hline \end{array}$

Do not forget to check your subtractions by addition. (See the method used in Section 2.2.)

4.3 Subtraction with the spike abacus

A situation such as 34_{five} take away 12_{five} is straightforward:

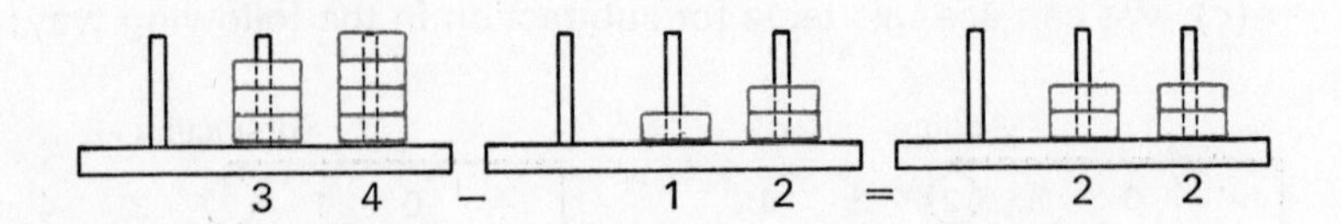

But, 31_{five} take away 3_{five} presents a problem:

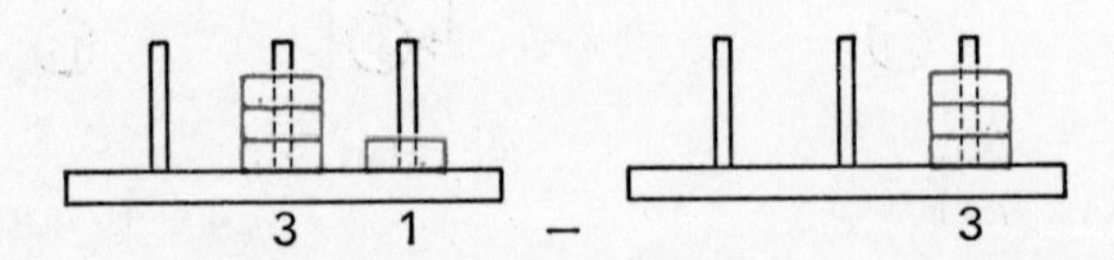

There is only one ring on the units spike and you want to remove three units. 'One take away three' cannot be done. To overcome this difficulty we have to use one of the rings on the fives spike and exchange it for five units:

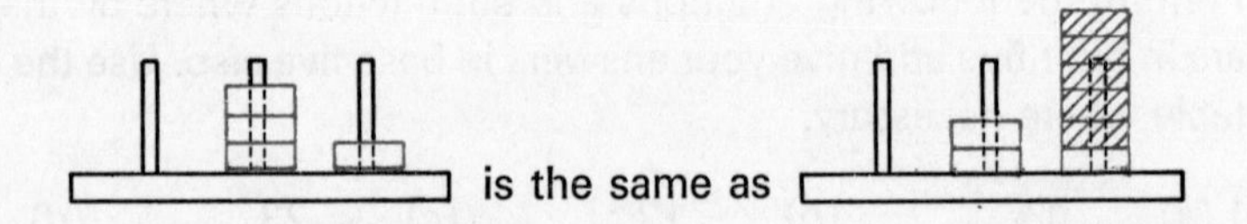

The five shaded rings represent the five-ring which has been removed. Now the problem is solved:

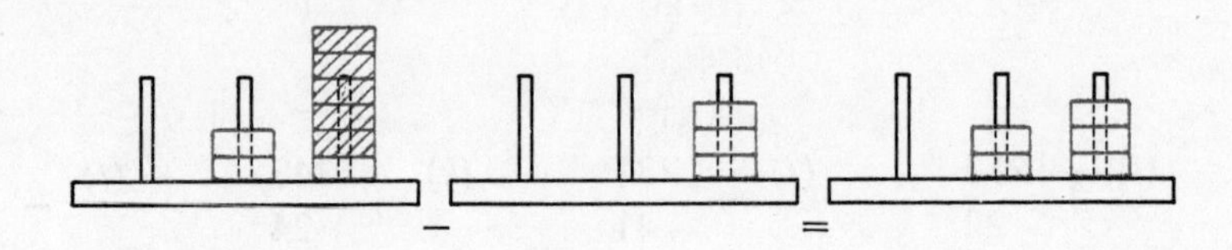

The answer is 23_{five}.

4.3 Subtraction with a spike abacus

Subtraction with the spike abacus focuses attention on the need for decomposition rather than other methods.

Exercise I

Note: the following abaci have been drawn with an extra spike; this is not introduced in the pupil's text until Section 4.5.

1 (*a*)

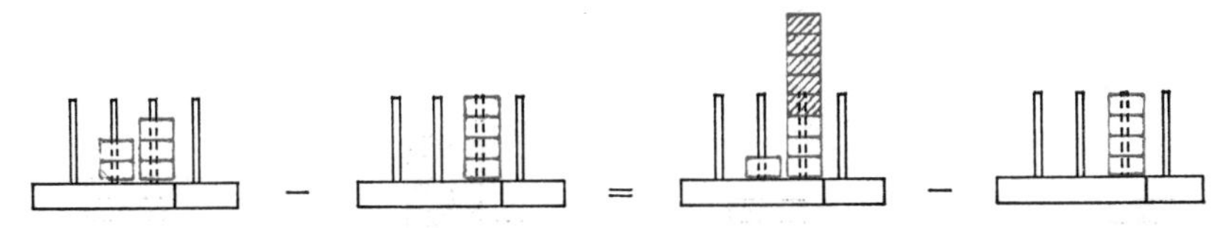

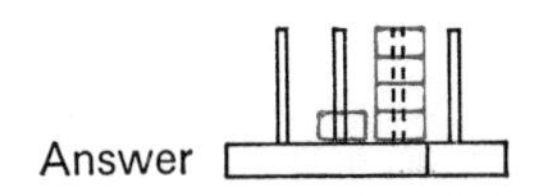

(*b*)

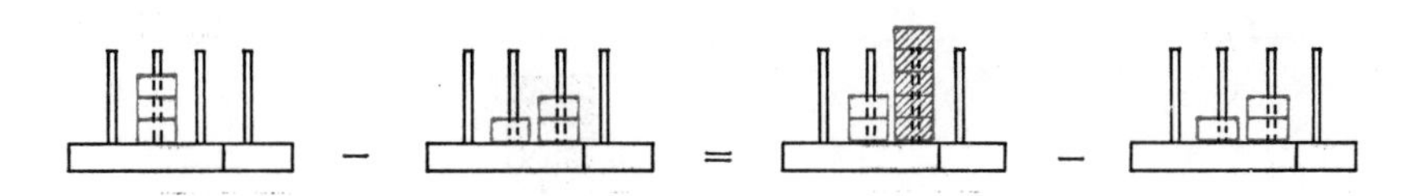

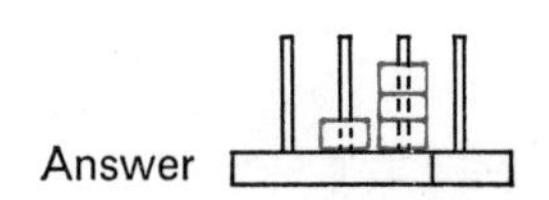

(*c*)

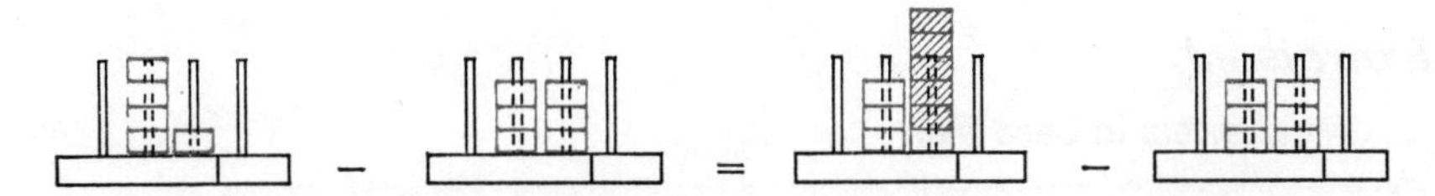

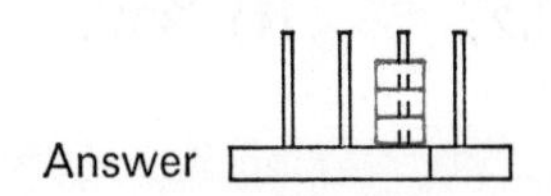

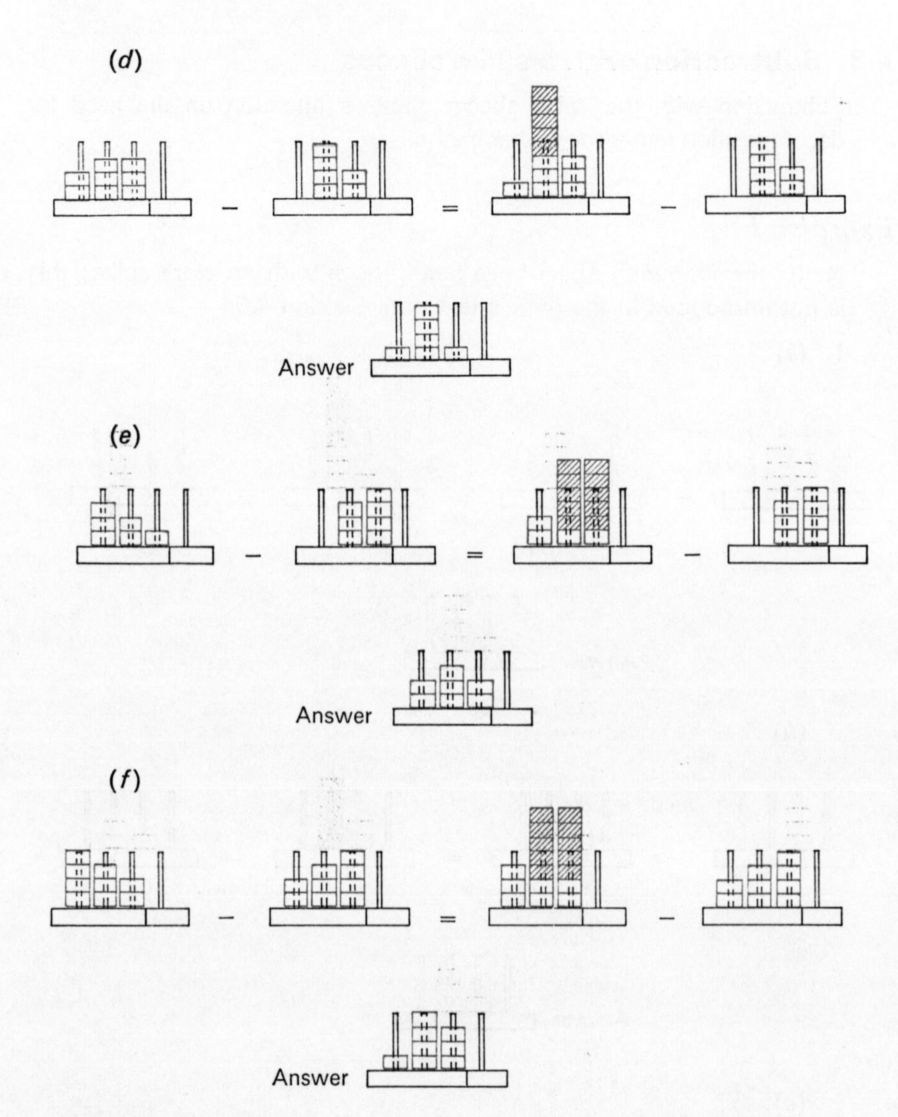

Exercise J

All answers in base five.

1 (*a*) 11; (*b*) 13; (*c*) 22; (*d*) 24;
(*e*) 31; (*f*) 113; (*g*) 103; (*h*) 124;
(*i*) 332; (*j*) 132; (*k*) 331; (*l*) 4242.

Exercise I

Draw spike abacus diagrams to represent the steps to be taken to carry out these base five subtractions.

1 (*a*) 23−4; (*b*) 30−12;

(*c*) 41−33; (*d*) 233−42;

(*e*) 321−34; (*f*) 432−234.

4.4 Multiplying in base five

Multiplying in base five should present no problems.

Exercise J

All these numbers are written in base five. Give the answers in base five.

1 (*a*) $\begin{array}{r} 3 \\ \times\ 2 \\ \hline \\ \hline \end{array}$ (*b*) $\begin{array}{r} 4 \\ \times\ 2 \\ \hline \\ \hline \end{array}$ (*c*) $\begin{array}{r} 4 \\ \times\ 3 \\ \hline \\ \hline \end{array}$ (*d*) $\begin{array}{r} 12 \\ \times\ 2 \\ \hline \\ \hline \end{array}$

(*e*) $\begin{array}{r} 13 \\ \times\ 2 \\ \hline \\ \hline \end{array}$ (*f*) $\begin{array}{r} 21 \\ \times\ 3 \\ \hline \\ \hline \end{array}$ (*g*) $\begin{array}{r} 12 \\ \times\ 4 \\ \hline \\ \hline \end{array}$ (*h*) $\begin{array}{r} 23 \\ \times\ 3 \\ \hline \\ \hline \end{array}$

(*i*) $\begin{array}{r} 43 \\ \times\ 4 \\ \hline \\ \hline \end{array}$ (*j*) $\begin{array}{r} 12 \\ \times\ 11 \\ \hline \\ \hline \end{array}$ (*k*) $\begin{array}{r} 23 \\ \times\ 12 \\ \hline \\ \hline \end{array}$ (*l*) $\begin{array}{r} 134 \\ \times\ 23 \\ \hline \\ \hline \end{array}$

Copy this multiplication table into your book and complete it. A few of the entries have already been made, to start you off.

×	0	1	2	3	4
0	0	0	0		
1					
2					
3	0			14	
4			13		

Look for patterns

The multiplication table can be used to give answers to division questions just as the addition table helped with subtraction.

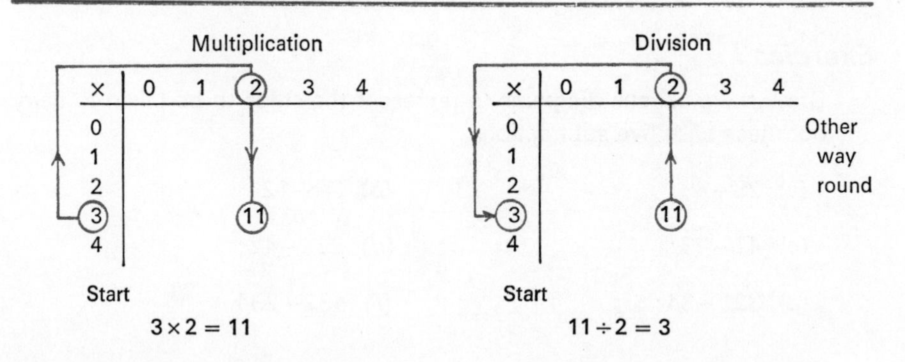

Exercise K

All these numbers are in base five. Give the answers in base five.

1 (*a*) $2\overline{)13}$ (*b*) $4\overline{)22}$ (*c*) $4\overline{)13}$

(*d*) $4\overline{)31}$ (*e*) $2\overline{)11}$ (*f*) $3\overline{)413}$

4.5 Using an extra spike

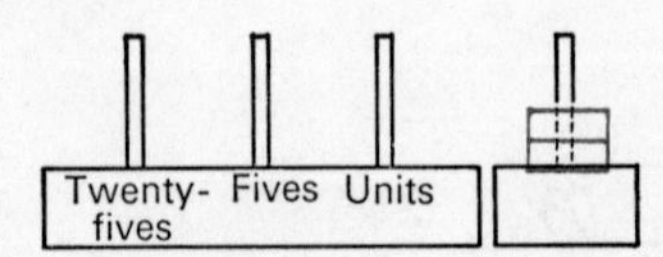

Start counting with the spike abacus by putting rings on the extra spike first. This time you will have to use five rings to get the position below.

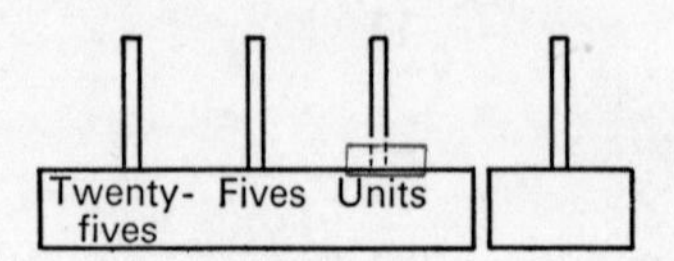

We have called the right-hand spike the 'extra one'. What is a better name for it? Remember, five on this spike are needed to make a whole one.

Call this the 'one-fifth' spike; rings placed upon it each represent the fraction one-fifth.

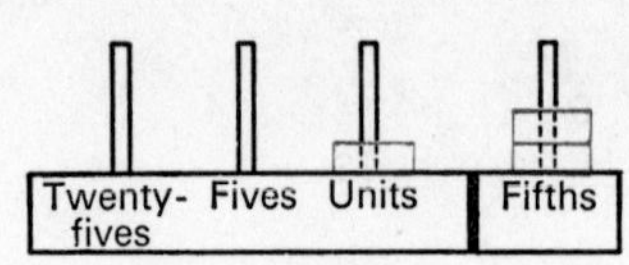

This situation represents one and two-fifths.

A note on the multiplication table in base five:

×	0	1	2	3	4
0	0	0	0	0	0
1	0	1	2	3	4
2	0	2	4	11	13
3	0	3	11	14	22
4	0	4	13	22	31

The direct use of the table for the multiplication $3 \times 4 = 22$ is shown.

If one now considers an inverse process for dividing:

×	0	1	2	3	4
0	0	0	0	0	0
1	0	1	2	3	4
2	0	2	4	11	13
3	0	3	11	14	22
4	0	4	13	22	31

The 'inverse' use of the table for the division $22 \div 4 = 3$ is shown.

(Notice that the same 'inverse' use of the table would suggest that

$$0 \div 0 = 4, \quad 0 \div 0 = 3, \quad 0 \div 0 = 2$$

and so on. There is obviously no particular value for $0 \div 0$. Multiplication by 0 is all right, but division by 0 is not.)

Exercise K

All answers in base five.

1 (*a*) 4; (*b*) 3; (*c*) 2; (*d*) 4; (*e*) 3; (*f*) 121.

4.5 Using an extra spike

If all the preceding work has been absorbed, there should be no difficulty here. Care should be taken not to overstress this situation or to suggest that anything unusual has happened. It follows naturally if the idea of place value has been grasped.

Exercise L

Answers in base ten.

1 $3\frac{1}{5}$. 2 $8\frac{1}{5}$. 3 $10\frac{3}{5}$.

4 (*a*) (*b*) (*c*)

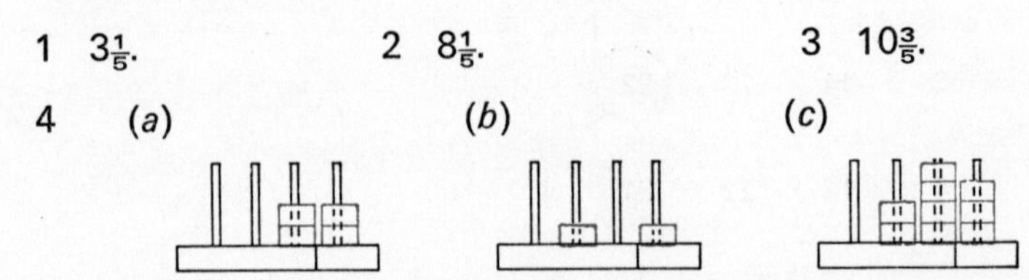

5 (*a*) $4\frac{3}{5}$; (*b*) $4\frac{2}{5}$; (*c*) $7\frac{2}{5}$; (*d*) $\frac{4}{5}$; (*e*) $3\frac{4}{5}$; (*f*) $1\frac{3}{5}$.

4.6 The fraction point

In base ten we can speak of the decimal point, but this is not so in other bases. Again, attempts have been made to describe this situation in other bases, such as 'bicimal point' for base two. The whole difficulty can be overcome by using 'fraction point' for all of them, including even base ten. This also emphasizes that we are dealing with fractions.

Exercise M

Answers to Question 1 are in base five.

1 (*a*) 4·1; (*b*) 3·2; (*c*) 1·4;
(*d*) 1·1; (*e*) 2·0; (*f*) 0·3.

Exercise L

The questions of this exercise are written in base ten. Give the answers in base ten.

What numbers are represented by these situations?

1 2 3

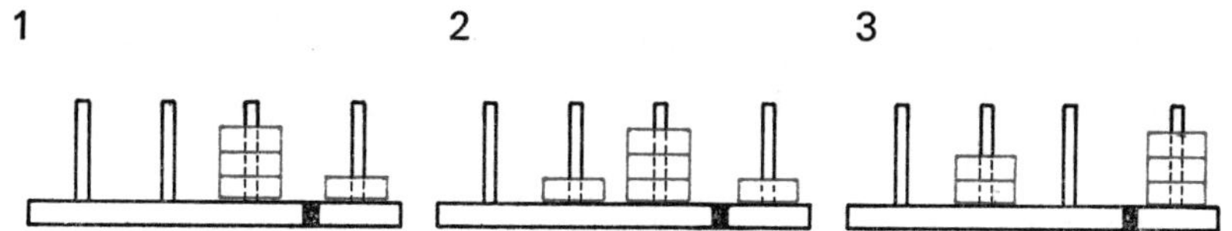

Draw diagrams to illustrate the positions on a base five abacus for:

4 (*a*) $2\frac{2}{5}$; (*b*) $5\frac{1}{5}$; (*c*) $14\frac{3}{5}$.

Use your spike abacus to do these:

5 (*a*) $3\frac{1}{5}+1\frac{2}{5}$; (*b*) $2\frac{3}{5}+1\frac{4}{5}$; (*c*) $4\frac{4}{5}+2\frac{3}{5}$;

(*d*) $2\frac{2}{5}-3\frac{1}{5}$; (*e*) $4\frac{1}{5}-\frac{2}{5}$; (*f*) $3\frac{2}{5}-1\frac{4}{5}$.

4.6 The fraction point

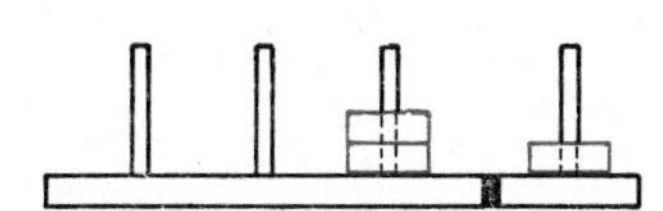

The number shown on this abacus is $2\frac{1}{5}$. This can be more neatly written as $2{\cdot}1_{\text{five}}$ using a dot to represent the band on the abacus. The dot is placed level with the middle of the digits.

Questions 1–3 of Exercise L can now be written as

$3{\cdot}1_{\text{five}}$, $13{\cdot}1_{\text{five}}$, $20{\cdot}3_{\text{five}}$.

The dot is read 'point' and a numeral such as 4·1 would be read 'four point one'. The dot is sometimes called a 'fraction point'.

Exercise M

These numbers are in base ten. Write them in base five using a fraction point.

1 (*a*) $4\frac{1}{5}$; (*b*) $3\frac{2}{5}$; (*c*) $1\frac{4}{5}$;

(*d*) $1\frac{1}{5}$; (*e*) 2; (*f*) $\frac{3}{5}$.

Now work the other way round. These numbers are in base five, put them back to base ten writing out the fractions:

2 (*a*) 3·2; (*b*) 4·1; (*c*) 3·3;
(*d*) 2·0; (*e*) 1·4; (*f*) 0·2.

The remaining questions in this exercise are written in base five. Add, and give your answers in base five with the fraction point *and* in base ten with a fraction:

3 (*a*) 2·1+1·2; (*b*) 4·3+0·1; (*c*) 1·3+2·1.

Subtract, and give your answers in base five with the fraction point and in base ten with a fraction:

4 (*a*) 3·4−2·3; (*b*) 0·4−0·1; (*c*) 4·4−3·0.

Add and give your answer in the fraction point method.

5 (*a*) $\begin{array}{r} 2{\cdot}2 \\ +\ 1{\cdot}4 \\ \hline \\ \hline \end{array}$ (*b*) $\begin{array}{r} 2{\cdot}3 \\ +\ 1{\cdot}3 \\ \hline \\ \hline \end{array}$ (*c*) $\begin{array}{r} 3{\cdot}3 \\ +\ 3{\cdot}4 \\ \hline \\ \hline \end{array}$

(*d*) $\begin{array}{r} 1{\cdot}4 \\ +\ 4{\cdot}1 \\ \hline \\ \hline \end{array}$ (*e*) $\begin{array}{r} 13{\cdot}2 \\ +\ 3{\cdot}1 \\ \hline \\ \hline \end{array}$ (*f*) $\begin{array}{r} 20{\cdot}3 \\ +\ 41{\cdot}4 \\ \hline \\ \hline \end{array}$

Subtract and give your answer in the fraction point method.

6 (*a*) $\begin{array}{r} 3{\cdot}1 \\ -\ 2{\cdot}1 \\ \hline \\ \hline \end{array}$ (*b*) $\begin{array}{r} 4{\cdot}3 \\ -\ 2{\cdot}4 \\ \hline \\ \hline \end{array}$ (*c*) $\begin{array}{r} 32{\cdot}3 \\ -\ 4{\cdot}1 \\ \hline \\ \hline \end{array}$

(*d*) $\begin{array}{r} 21{\cdot}2 \\ -\ 3{\cdot}0 \\ \hline \\ \hline \end{array}$ (*e*) $\begin{array}{r} 231{\cdot}3 \\ -\ 40{\cdot}1 \\ \hline \\ \hline \end{array}$ (*f*) $\begin{array}{r} 320{\cdot}3 \\ -\ 11{\cdot}2 \\ \hline \\ \hline \end{array}$

Did you check all your subtractions by the method we used before?

4.7 Bases four and six

There was no special reason for choosing base five. If you think of the original idea of grouping which we used, you will see that we can count in any number as a base.

Answers to Questions 2 are in base ten.

2 (*a*) $3\frac{2}{5}$; (*b*) $4\frac{1}{5}$; (*c*) $3\frac{3}{5}$;
(*d*) 2; (*e*) $1\frac{4}{5}$; (*f*) $\frac{2}{5}$.

3 (*a*) $3{\cdot}3_{\text{five}}$, $3\frac{3}{5}_{\text{ten}}$; (*b*) $4{\cdot}4_{\text{five}}$, $4\frac{4}{5}_{\text{ten}}$; (*c*) $3{\cdot}4_{\text{five}}$, $3\frac{4}{5}_{\text{ten}}$.

4 (*a*) $1{\cdot}1_{\text{five}}$, $1\frac{1}{5}_{\text{ten}}$; (*b*) $0{\cdot}3_{\text{five}}$, $\frac{3}{5}_{\text{ten}}$; (*c*) $1{\cdot}4_{\text{five}}$, $1\frac{4}{5}_{\text{ten}}$.

Answers to Questions 5 and 6 are in base five.

5 (*a*) 4·1; (*b*) 4·1; (*c*) 12·2;
(*d*) 11·0; (*e*) 21·3; (*f*) 112·2.

6 (*a*) 1·0; (*b*) 1·4; (*c*) 23·2;
(*d*) 13·2; (*e*) 141·2; (*f*) 304·1.

4.7 Bases four and six

This section and Exercise N can be omitted without much loss.

Number bases

Exercise N

1

	Base ten	Base four	Base five	Base six
a	7	13	12	11
b	10	22	20	14
c	12	30	22	20
d	29	131	104	45
e	18	102	33	30
f	52	310	202	124
g	33	201	113	53
h	41	221	131	105
i	51	303	201	123
j	43	223	133	111

2

+	0	1	2	3
0	0	1	2	3
1	1	2	3	10
2	2	3	10	11
3	3	10	11	12

×	0	1	2	3
0	0	0	0	0
1	0	1	2	3
2	0	2	10	12
3	0	3	12	21

3

+	0	1	2	3	4	5
0	0	1	2	3	4	5
1	1	2	3	4	5	10
2	2	3	4	5	10	11
3	3	4	5	10	11	12
4	4	5	10	11	12	13
5	5	10	11	12	13	14

×	0	1	2	3	4	5
0	0	0	0	0	0	0
1	0	1	2	3	4	5
2	0	2	4	10	12	14
3	0	3	10	13	20	23
4	0	4	12	20	24	32
5	0	5	14	23	32	41

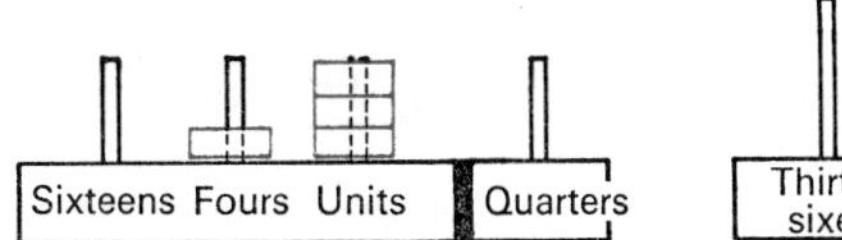

A base four abacus
showing
13_{four}

What is this in base ten?

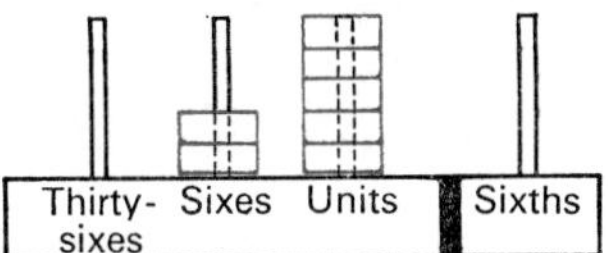

A base six abacus
showing
25_{six}

What is this in base ten?

You can easily alter your base five abacus to these or any other bases by pulling out the spikes and replacing them with ones of the desired height.

Exercise N

1 Copy and complete this table.

	Base ten	Base four	Base five	Base six
a	7			
b	10			
c		30		
d	29	131	104	45
e			33	
f			202	
g				53
h				105
i		303		
j				111

When in doubt refer to your abacus

2 Copy and complete these tables in base four.

+	0	1	2	3
0				
1				
2				
3				

×	0	1	2	3
0				
1				
2				
3				

3 Copy and complete these tables in base six.

+	0	1	2	3	4	5
0						
1						
2						
3						
4						
5						

×	0	1	2	3	4	5
0						
1						
2						
3						
4						
5						

4 All in base four,

(*a*) $\begin{array}{r} 22 \\ +\ \ 2 \\ \hline \\ \hline \end{array}$ (*b*) $\begin{array}{r} 101 \\ +\ \ 23 \\ \hline \\ \hline \end{array}$ (*c*) $\begin{array}{r} 23 \\ +\ \ 12 \\ \hline \\ \hline \end{array}$ (*d*) $\begin{array}{r} 32 \\ -\ \ 13 \\ \hline \\ \hline \end{array}$

(*e*) $\begin{array}{r} 123 \\ -\ \ 30 \\ \hline \\ \hline \end{array}$ (*f*) $\begin{array}{r} 12 \\ \times\ \ 2 \\ \hline \\ \hline \end{array}$ (*g*) $\begin{array}{r} 203 \\ \times\ \ 3 \\ \hline \\ \hline \end{array}$ (*h*) $\begin{array}{r} 332 \\ \times\ \ 12 \\ \hline \\ \hline \end{array}$

5 All in base six,

(*a*) $\begin{array}{r} 34 \\ +\ \ 13 \\ \hline \\ \hline \end{array}$ (*b*) $\begin{array}{r} 205 \\ +\ \ 24 \\ \hline \\ \hline \end{array}$ (*c*) $\begin{array}{r} 333 \\ +\ \ 34 \\ \hline \\ \hline \end{array}$ (*d*) $\begin{array}{r} 44 \\ -\ \ 5 \\ \hline \\ \hline \end{array}$

(*e*) $\begin{array}{r} 320 \\ -\ \ 24 \\ \hline \\ \hline \end{array}$ (*f*) $\begin{array}{r} 12 \\ \times\ \ 3 \\ \hline \\ \hline \end{array}$ (*g*) $\begin{array}{r} 421 \\ \times\ \ 4 \\ \hline \\ \hline \end{array}$ (*h*) $\begin{array}{r} 224 \\ \times\ \ 21 \\ \hline \\ \hline \end{array}$

4.8 Using fixed scales to add and subtract

Here is another device to help you work in base five:

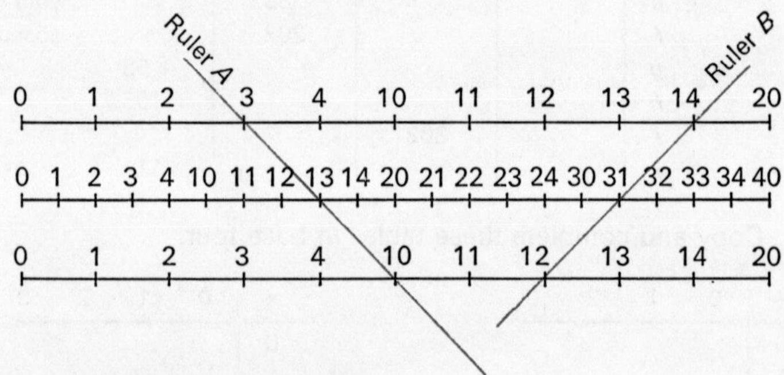

Copy these scales into your book

To add two numbers together, find one number on the top scale and the other on the bottom. Lay a ruler across and the answer will appear on the middle scale.

Ruler *A* shows: $3_{\text{five}}+10_{\text{five}}=13_{\text{five}}$,

Ruler *B* shows: $14_{\text{five}}+12_{\text{five}}=31_{\text{five}}$.

Exercise O

1 Perform these base five additions with the aid of the adding scales:

(*a*) 13+14; (*b*) 2+11; (*c*) 4+13; (*d*) 14+14.

4 All answers in base four:

(*a*) 30; (*b*) 130; (*c*) 101; (*d*) 13;
(*e*) 33; (*f*) 30; (*g*) 1221; (*h*) 11310.

5 All answers in base six:

(*a*) 51; (*b*) 233; (*c*) 411; (*d*) 35;
(*e*) 252; (*f*) 40; (*g*) 2524; (*h*) 5144.

4.8 Using fixed scales to add and subtract

This is an example of a nomogram; nomograms are used by engineers, but are included here as a simple practical task. Some pupils might like to put these scales on card and even make other nomograms to perform addition in other bases.

Exercise O

Answers to Questions 1 and 2 are in base five.

1 (*a*) 32; (*b*) 13; (*c*) 22; (*d*) 33.

2 (*a*) 3; (*b*) 3; (*c*) 11.

3

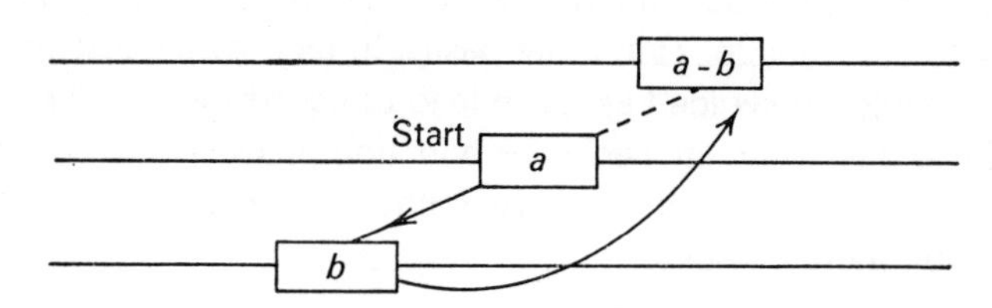

4 Lines must be parallel and equidistant. The two outer scales must be the same and the middle one has to be double the outer ones.

(*a*) Base four:

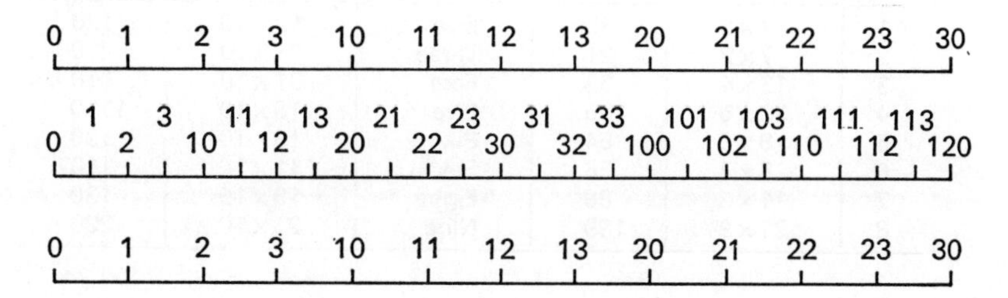

(*b*) Base six:

0 1 2 3 4 5 10 11 12 13 14 15 20

0 1 2 3 4 5 10 11 12 13 14 15 20 21 22 23 24 25 30 31 32 33 34 35 40

0 1 2 3 4 5 10 11 12 13 14 15 20

(*c*) Base ten:

0 1 2 3 4 5 6 7 8 9 10 11 12

0 1 2 3 4 5 6 7 8 9 10 11 12 13 14 15 16 17 18 19 20 21 22 23 24

0 1 2 3 4 5 6 7 8 9 10 11 12

4.9 Multiplication by the base

This is really quite important. The rule of 'adding a nought' to multiply by ten in base ten still applies in other bases. In fact the expression 'adding a nought' should be avoided as it can lead to confusion. It is better to think of all the digits being moved over and noughts used to fill the empty spaces; consideration of this in bases other than ten should help to make the situation more obvious.

Exercise P

	Question (base ten)	Answer	Base	Question	Answer
1	7×5	35	Five	12×10	120
2	7×3	21	Three	21×10	210
3	13×4	52	Four	31×10	310
4	33×5	165	Five	113×10	1130
5	9×6	54	Six	13×10	130
6	8×7	56	Seven	11×10	110
7	11×8	88	Eight	13×10	130
8	21×9	189	Nine	23×10	230

2 See if you can 'work backwards' and use the scales to do these:

(*a*) 4−1; (*b*) 12−4; (*c*) 30−14.

3 Explain how you would use the scales to subtract.

4 Study the scales and see if you can see how they were made. See if you can make other scales which will work in:

(*a*) base four; (*b*) base six; (*c*) base ten.

4.9 Multiplication by the base

In everyday base ten arithmetic, to multiply by ten is very simple. Multiplication by ten just moves every number over to the next place on the left and we put a nought to fill the empty space.

For example:

h	t	u		t	u		h	t	u
	1	7	×	1	0	=	1	7	0

Will there be the same sort of rule in other bases?
What happens if you multiply a base five number by five?
What happens if you multiply a base eight number by eight?
Exercise P will help you answer these questions.

Exercise P

First do the question in base ten, then change it into the given base and do it again.

The first two have been done for you as examples:

	Question (base ten)	Answer	Base	Question	Answer
	23 × 5	115	Five	43 × 10	430
	11 × 7	77	Seven	14 × 10	140
1	7 × 5		Five		
2	7 × 3		Three		
3	13 × 4		Four		
4	33 × 5		Five		
5	9 × 6		Six		
6	8 × 7		Seven		
7	11 × 8		Eight		
8	21 × 9		Nine		

Can you now make up a general rule which applies whenever you multiply a number, written in a certain base, by that base?

Miscellaneous Exercise Q

1 Add: 5 weeks 3 days, 1 week 2 days and 1 week 6 days.

2 Add: 2 years 5 months, 1 year 8 months and 7 months.

3 Put in column headings to make these correct. (Use imperial units.)

(*a*) $\begin{array}{r} 25 \\ +\;16 \\ \hline 43 \\ \hline \end{array}$ (*b*) $\begin{array}{r} 41 \\ +\;21 \\ \hline 70 \\ \hline \end{array}$ (*c*) $\begin{array}{r} 12 \\ +\;12 \\ \hline 31 \\ \hline \end{array}$ (*d*) $\begin{array}{r} 39 \\ +\;19 \\ \hline 52 \\ \hline \end{array}$

(*e*) $\begin{array}{r} 28 \\ +\;18 \\ \hline 42 \\ \hline \end{array}$ (*f*) $\begin{array}{r} 35 \\ +\;34 \\ \hline 72 \\ \hline \end{array}$ (*g*) $\begin{array}{r} 13 \\ \times\;3 \\ \hline 42 \\ \hline \end{array}$ (*h*) $\begin{array}{r} 13 \\ \times\;3 \\ \hline 51 \\ \hline \end{array}$

(*i*) $\begin{array}{r} 13 \\ \times\;3 \\ \hline 71 \\ \hline \end{array}$ (*j*) $\begin{array}{r} 64 \\ -\;58 \\ \hline 8 \\ \hline \end{array}$ (*k*) $\begin{array}{r} 71 \\ -\;42 \\ \hline 22 \\ \hline \end{array}$ (*l*) $\begin{array}{r} 22 \\ -\;9 \\ \hline 19 \\ \hline \end{array}$

4 Count these dots in the usual base ten manner, then group into sixes and represent the situation in number symbols to base six.

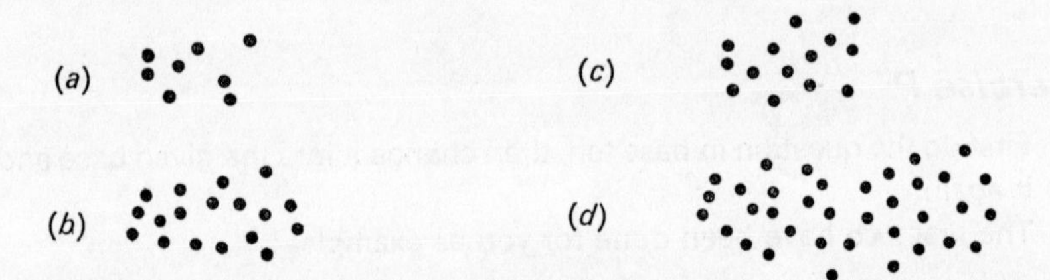

5 Write down in words the meaning of 40_{six}. Why is the 0 there?

6 Have you noticed that in base five only the symbols, or digits, 0, 1, 2, 3 and 4 are used? What are the digits we can use in:

(*a*) base four; (*b*) base seven; (*c*) base nine; (*d*) base ten?

7 What does the 4 mean in 43 if it is in:

(*a*) base ten; (*b*) base six; (*c*) base eight?

What is the lowest base 43 can be written in?

8 Write down the number of these dots

(*a*) in base ten; (*b*) in base four; (*c*) in base five.

Miscellaneous Exercise Q

1 8 weeks, 4 days.

2 4 years, 8 months.

3 (*a*) Gal and pt; (*b*) qt and pt; (*c*) yd and ft;
(*d*) lb and oz; (*e*) st and lb; (*f*) weeks and days;
(*g*) weeks and days; (*h*) gal and qt; (*i*) qt and pt;
(*j*) ft and in.; years and months; (*k*) yd and ft;
(*l*) lb and oz.

4 (*a*) 12_{six}; (*b*) 30_{six}; (*c*) 22_{six}; (*d*) 100_{six}.

5 Four sixes and no units. The zero is used to fill the empty units space.

6 (*a*) Base four uses 0, 1, 2, 3;
(*b*) base seven uses 0, 1, 2, 3, 4, 5, 6;
(*c*) base nine uses 0, 1, 2, 3, 4, 5, 6, 7, 8;
(*d*) base ten uses 0, 1, 2, 3, 4, 5, 6, 7, 8, 9.

7 (*a*) Four tens; (*b*) four sixes; (*c*) four eights.
43 could not be written in a base lower than five.

8 (*a*) 12; (*b*) 30; (*c*) 22.

Number bases

9

+	0	1	2	3	4	5	6	7
0	0	1	2	3	4	5	6	7
1	1	2	3	4	5	6	7	10
2	2	3	4	5	6	7	10	11
3	3	4	5	6	7	10	11	12
4	4	5	6	7	10	11	12	13
5	5	6	7	10	11	12	13	14
6	6	7	10	11	12	13	14	15
7	7	10	11	12	13	14	15	16

×	0	1	2	3	4	5	6	7
0	0	0	0	0	0	0	0	0
1	0	1	2	3	4	5	6	7
2	0	2	4	6	10	12	14	16
3	0	3	6	11	14	17	22	25
4	0	4	10	14	20	24	30	34
5	0	5	12	17	24	31	36	43
6	0	6	14	22	30	36	44	52
7	0	7	16	25	34	43	52	61

10 All answers in base eight:

(*a*) 41; (*b*) 72; (*c*) 45; (*d*) 173;
(*e*) 270; (*f*) 415; (*g*) 25; (*h*) 36;
(*i*) 25; (*j*) 62; (*k*) 70; (*l*) 114.

11

	Base ten	Base five	Base six	Base eight
a	9	14	13	11
b	12	22	20	14
c	20	40	32	24
d	10	20	14	12
e	13	23	21	15
f	16	31	24	20
g	45	140	113	55
h	73	243	201	111
i	28	103	44	34
j	81	311	213	121

12 (*a*) $2\frac{1}{4}$; (*b*) $3\frac{3}{4}$; (*c*) $\frac{2}{4}$;
(*d*) $3\frac{1}{6}$; (*e*) $1\frac{5}{6}$; (*f*) $\frac{3}{6}$;
(*g*) $2\frac{7}{8}$; (*h*) $4\frac{5}{8}$; (*i*) $\frac{3}{8}$.

9 Copy and complete these addition and multiplication squares for base eight.

+	0	1	2	3	4	5	6	7
0								
1								
2								
3								
4								
5								
6								
7								

×	0	1	2	3	4	5	6	7
0								
1								
2								
3								
4								
5								
6								
7								

10 All in base eight.

(*a*) $\begin{array}{r} 27 \\ +\ 12 \\ \hline \\ \hline \end{array}$ (*b*) $\begin{array}{r} 45 \\ +\ 25 \\ \hline \\ \hline \end{array}$ (*c*) $\begin{array}{r} 26 \\ +\ 17 \\ \hline \\ \hline \end{array}$ (*d*) $\begin{array}{r} 127 \\ +\ 44 \\ \hline \\ \hline \end{array}$

(*e*) $\begin{array}{r} 205 \\ +\ 63 \\ \hline \\ \hline \end{array}$ (*f*) $\begin{array}{r} 343 \\ +\ 52 \\ \hline \\ \hline \end{array}$ (*g*) $\begin{array}{r} 32 \\ -\ 5 \\ \hline \\ \hline \end{array}$ (*h*) $\begin{array}{r} 52 \\ -\ 14 \\ \hline \\ \hline \end{array}$

(*i*) $\begin{array}{r} 74 \\ -\ 47 \\ \hline \\ \hline \end{array}$ (*j*) $\begin{array}{r} 12 \\ \times\ 5 \\ \hline \\ \hline \end{array}$ (*k*) $\begin{array}{r} 34 \\ \times\ 2 \\ \hline \\ \hline \end{array}$ (*l*) $\begin{array}{r} 46 \\ \times\ 2 \\ \hline \\ \hline \end{array}$

11 Copy and complete:

	Base ten	Base five	Base six	Base eight
a	9			
b	12			
c	20			
d		20		
e			21	
f				20
g		140		
h			201	
i				34
j				121

12 Write these in a different form:

(*a*) $2{\cdot}1_{\text{four}}$; (*b*) $3{\cdot}3_{\text{four}}$; (*c*) $0{\cdot}2_{\text{four}}$;

(*d*) $3{\cdot}1_{\text{six}}$; (*e*) $1{\cdot}5_{\text{six}}$; (*f*) $0{\cdot}3_{\text{six}}$;

(*g*) $2{\cdot}7_{\text{eight}}$; (*h*) $4{\cdot}5_{\text{eight}}$; (*i*) $0{\cdot}3_{\text{eight}}$.

13 The clues are in the bases as shown but the answers are to be written in base ten.

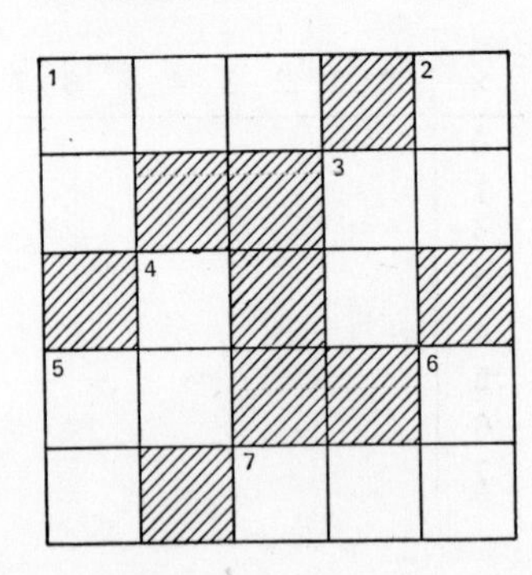

Across	Down
1. 411_{five}	1. 23_{four}
3. 33_{four}	2. 300_{five}
5. 42_{eight}	3. 15_{seven}
7. 300_{six}	4. 136_{eight}
	5. 40_{nine}
	6. 44_{six}

14 Copy and complete this set of adding scales for base eight.

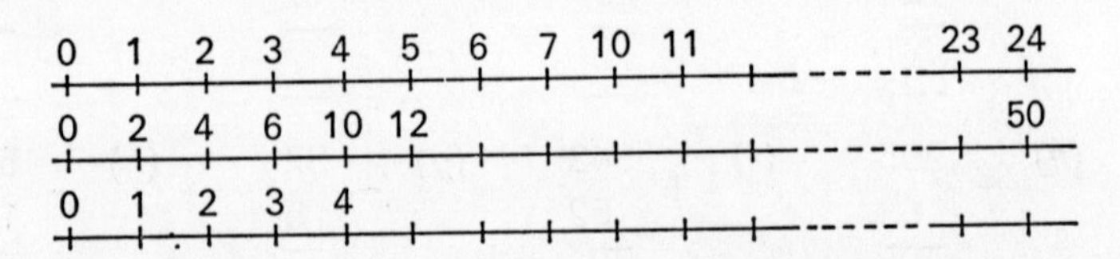

15 Use the scales you drew for Question 14 to answer these:

(*a*) $13_{eight}+17_{eight}$; (*b*) $23_{eight}+23_{eight}$;

(*c*) $46_{eight}-27_{eight}$.

13

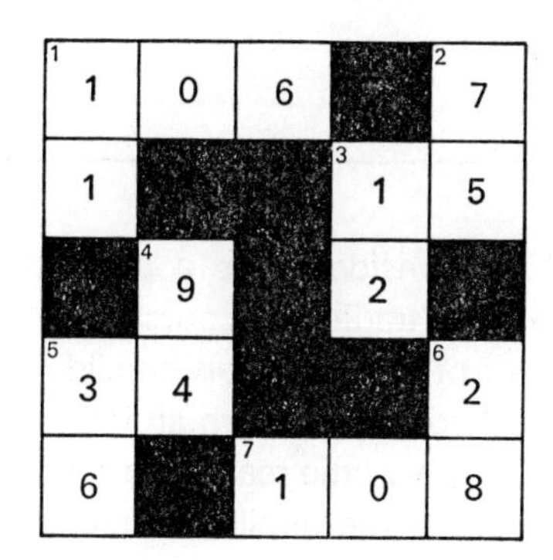

14

0 1 2 3 4 5 6 7 10 11 12 13 14 15 16 17 20 21 22 23 24

0 2 4 6 10 12 14 16 20 22 24 26 30 32 34 36 40 42 44 46 50

0 1 2 3 4 5 6 7 10 11 12 13 14 15 16 17 20 21 22 23 24

15 (*a*) 32_{eight}; (*b*) 46_{eight}; (*c*) 17_{eight}.

T

5. Symmetry

Symmetry could be considered as a property as fundamental as congruence has been in the traditional course. It would be possible to base the formal development of geometry on considerations of symmetry, but, in this course, we shall consider it from an informal point of view.

Formal proofs have very little real meaning for pupils at the beginning of the secondary stage, but the pupils can readily appreciate straightforward properties by considering the symmetry of figures. For example, the properties of an isosceles triangle will be obvious from considerations of symmetry and we should think it unnecessary to offer an apparently irrelevant 'proof'.

In this volume we have decided to introduce the topic by means of two-dimensional symmetry as it is in many ways easier for the pupils to create examples for themselves. In discussion there is, of course, no need to omit examples of three-dimensional symmetry.

1. INK DEVILS

The use of ink devils as an introductory section has three obvious advantages. It gives results which are clear and usually attractive in appearance; it is great fun for the pupils, and everybody can manage to produce a satisfactory result.

5. Symmetry

1. INK DEVILS

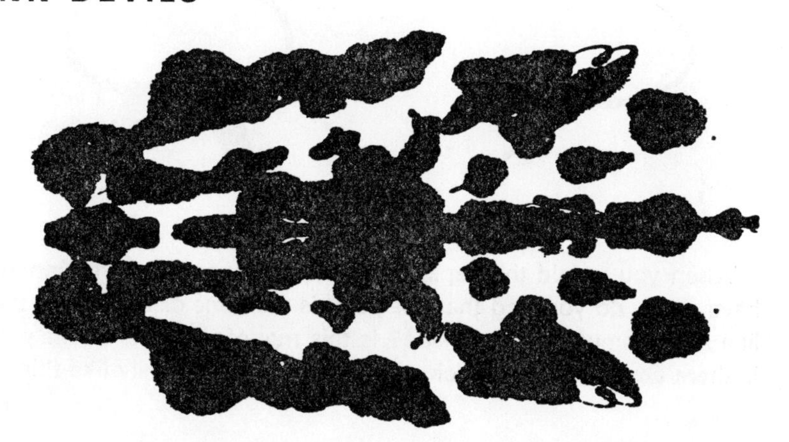

Many different patterns can be made using folded paper—sometimes the results are surprising but, if you are both quick and careful, you may be able to design a pattern.

Begin by folding a piece of paper; open it out again and scatter some blots of ink on it. Very big, wet blots will not make good devils so do not use too much! Fold the paper again on the same line as before; the ink will then spread out between the two layers of paper. When it is opened out again you will be able to see the pattern your blots have made.

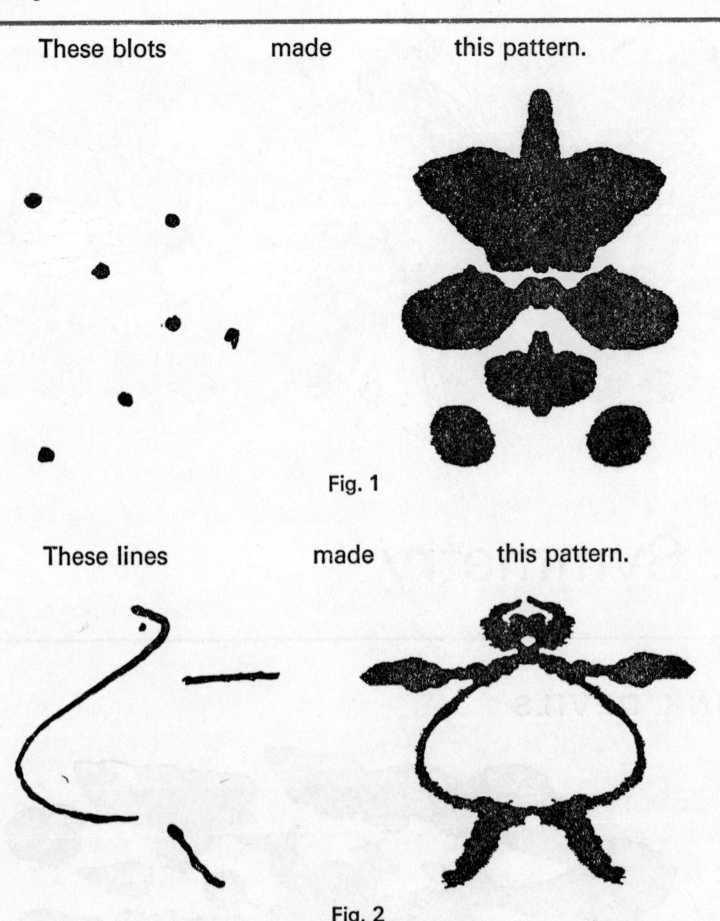

Fig. 1

Fig. 2

When you unfold the paper and look at any point of the pattern you have made, do you find that this point is the same distance from the fold line as the smudge it has made? Is this true of all points on the pattern? Is there any other line which divides the pattern equally like this?

2. LINE SYMMETRY

Any patterns like the ink-devils are called symmetrical (equal measure). The fold line is called the line of symmetry. Any point of the pattern will have its counterpart an equal distance on the opposite side of this line.

Exercise A

In Questions 1–5, the line of symmetry and half of the pattern have been drawn; copy them and complete the patterns.

2. LINE SYMMETRY

In two dimensions, the term 'line of symmetry' is preferred to 'axis of symmetry'. (The word 'axis' is retained for the axis of rotational symmetry in three dimensions.)

Symmetry

Exercise A

Before the class attempts this exercise it would be helpful if some examples were done on the blackboard, care being taken to see that the lines of symmetry are not all vertical or horizontal.

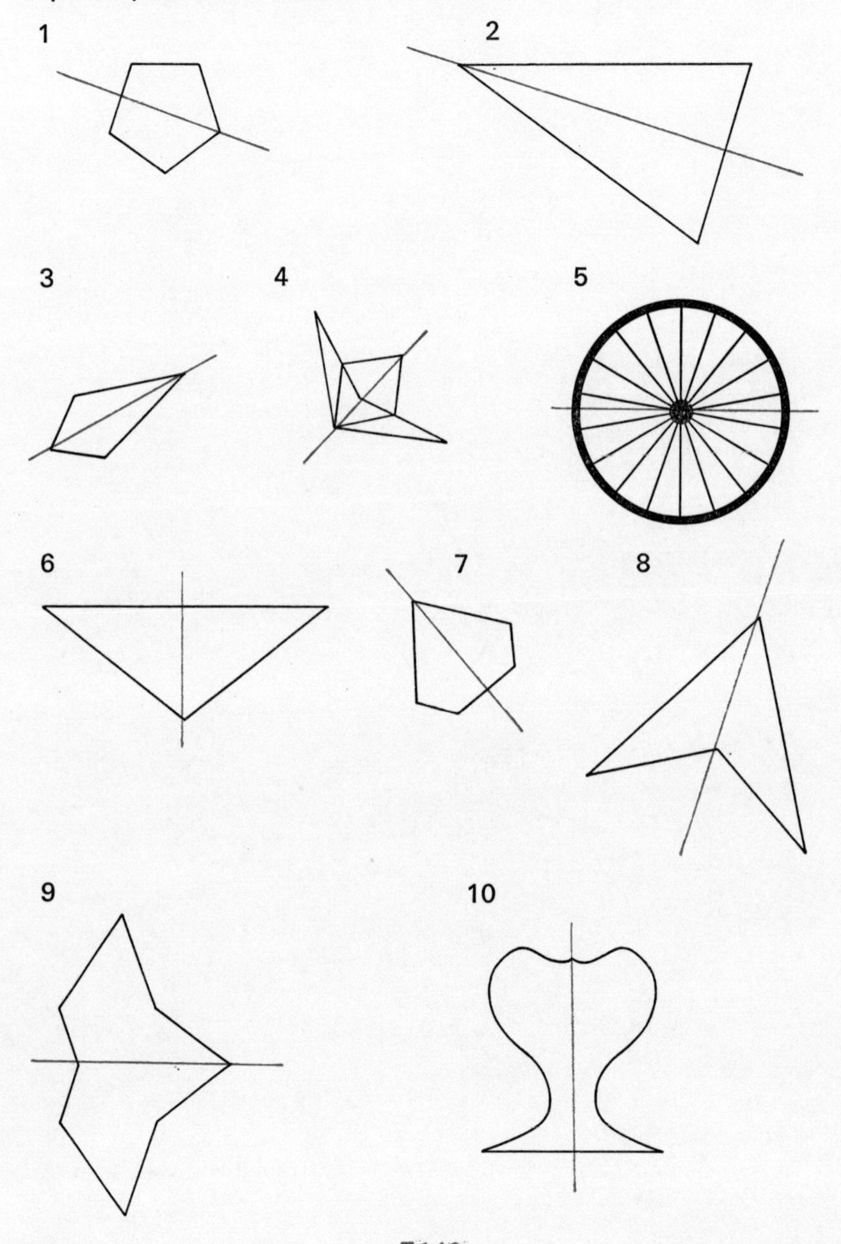

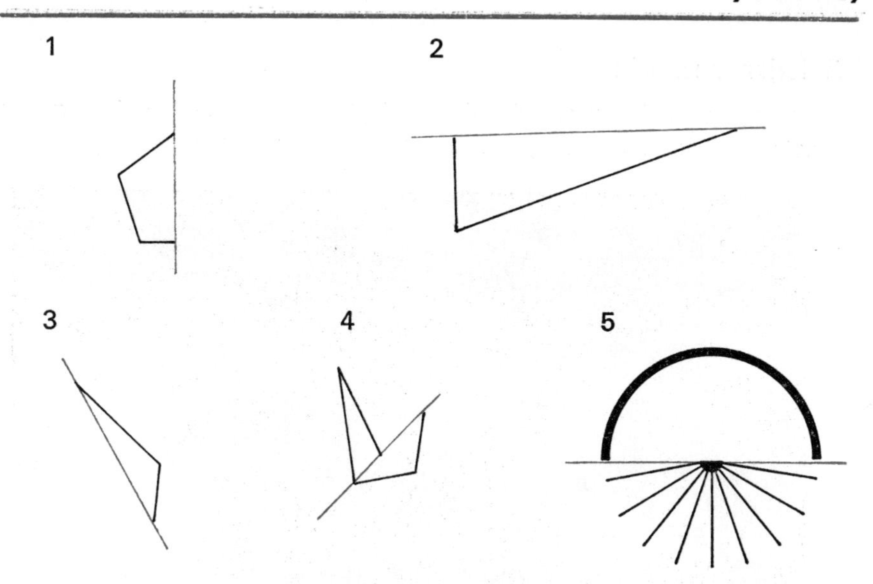

Work in pairs. One of you draw half a pattern as in Questions 1–5 and allow your neighbour to complete it.

Questions 6–10 show complete patterns. Copy them and draw in the line of symmetry in colour.

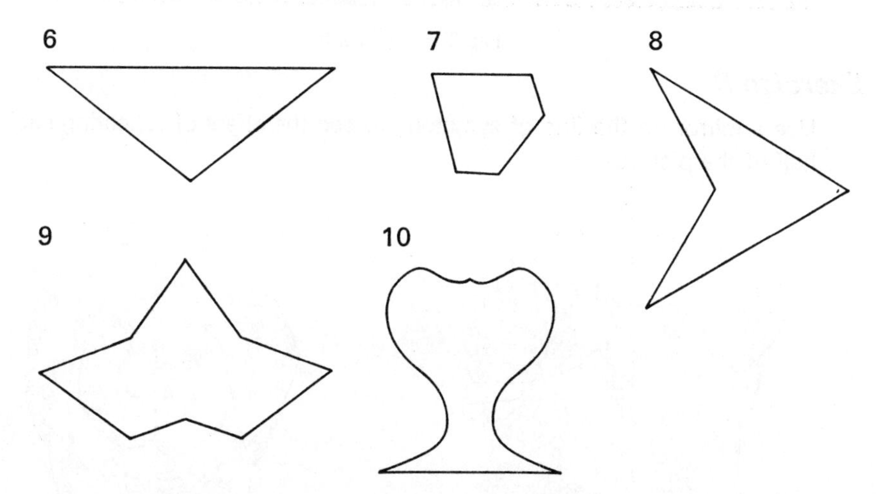

2.1 Ideas for everyone

Make a classroom display using:

(*a*) the best ink devils arranged with their lines of symmetry pointing in different directions;

(*b*) symmetrical parts of advertisements cut from magazines or newspapers. The brightly coloured ones are best.

2.2 Mirror magic

If you place a picture or a drawing against a mirror, the edge of the mirror acts as a line of symmetry between the picture and its reflection.

Fig. 3. Harry Worth

Exercise B

Use a mirror on the line of symmetry to see the effect of repeating each half of the picture.

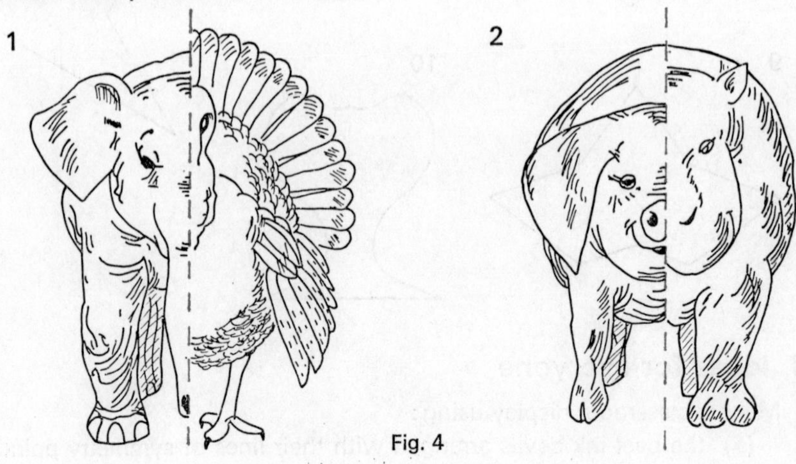

Fig. 4

3 Draw half-and-half objects as illustrated and use the mirror to complete them.

2.1 Ideas for everyone

Wall displays are very attractive and take very little time to construct. With the advertisements there will probably be a discussion about lettering, which often cuts across excellent pictures; if it is small it could be painted out. Mail order catalogues will also be found useful for gaily coloured pictures, as will pieces of wall paper.

2.2 Mirror magic

A certain amount of free play with mirrors often provides a good introduction. Attention can then be drawn to Questions 1–5 in Exercise A and the mirror placed on the line of symmetry.

When objects are used with a mirror there is a gap between the object and the image due to the thickness of the glass.

Exercise C

A large quantity of paper will be needed for this but thin paper (which is comparatively cheap) will be found to be quite adequate. Rolls of lining paper have been found to be useful.

1 (*a*) Triangle. (*b*) Triangle.
(*c*) One pair of sides is equal.
One pair of angles is equal.
(*d*) Line of symmetry.
Bisector of vertical angle.
Bisector of base.
(*e*) Equal.
(*f*) Cut at 45° to the fold.

2 (*a*) Triangle; (*b*) triangle.

2.3 More ideas for everyone

1 Collect as many different leaves as you can and press them between two pieces of paper weighted by heavy books. When they are really flat, find their lines of symmetry. Arrange and stick them on a large sheet of paper with the lines of symmetry marked in colour.

2 Make drawings or collect pictures of buildings or other man-made objects which show a clear line of symmetry.

Exercise C

You will need lots of paper to do this. Thin paper is cheaper and works very well.

1 Fold a piece of paper once (look at Figure 5), then cut off a corner.

(*a*) Write down the name of the shape you expect to cut off.
(*b*) Open out the piece you have cut off and write down the name of the shape you actually did get.
(*c*) What can you say about the sides and angles of this figure?
(*d*) What line does the fold represent?
(*e*) What can you say about the two parts of the figure on either side of the fold?
(*f*) Draw a diagram to show how you would cut the paper so that one of the angles of the figure would be a right-angle.

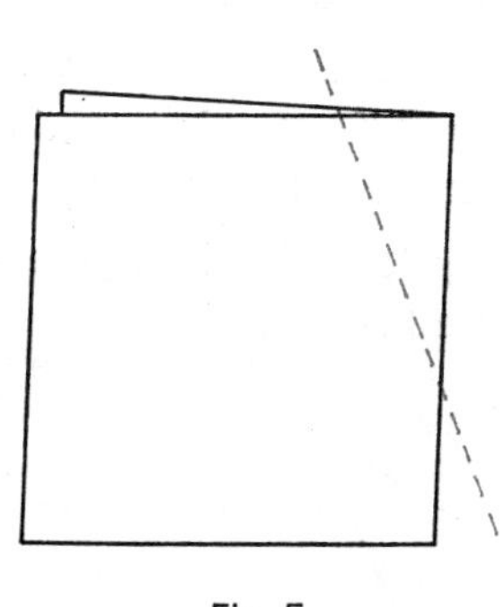

Fig. 5

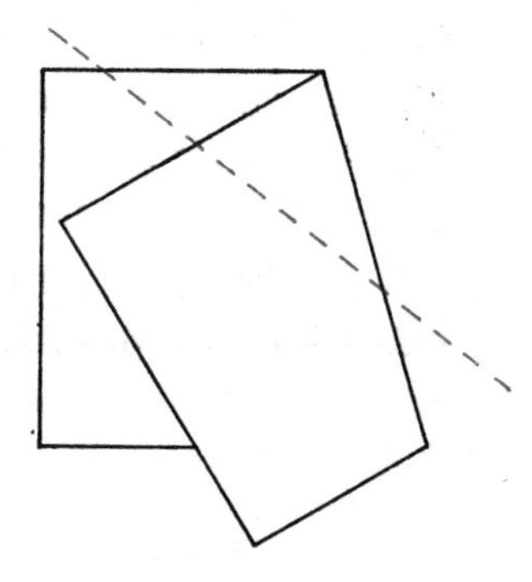

Fig. 6

2 Fold the paper once with an oblique fold (see Figure 6) and cut across the corner.

Write down:

(*a*) the name of the shape you expected to get;
(*b*) the name of the shape you do get.

3 Do the same as for Question 2 but make the oblique fold at a different angle. Answer the same two questions.

In your triangles for Questions 2 and 3 what does the fold line represent?

4 Fold a piece of paper in four as shown in Figure 7. Cut across the corner.

(*a*) What sort of shape did you expect to see this time?
(*b*) What shape have you actually got?
(*c*) What lines do the folds represent?
(*d*) Draw a diagram to show how you would make the cut so that the shape you got would be a square. Do the folds give all the lines of symmetry?

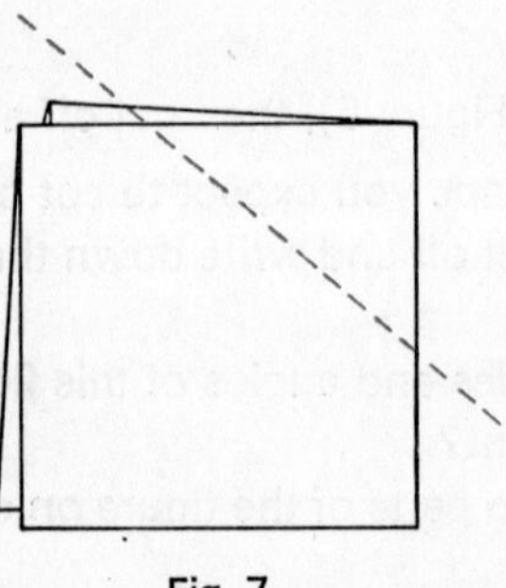

Fig. 7

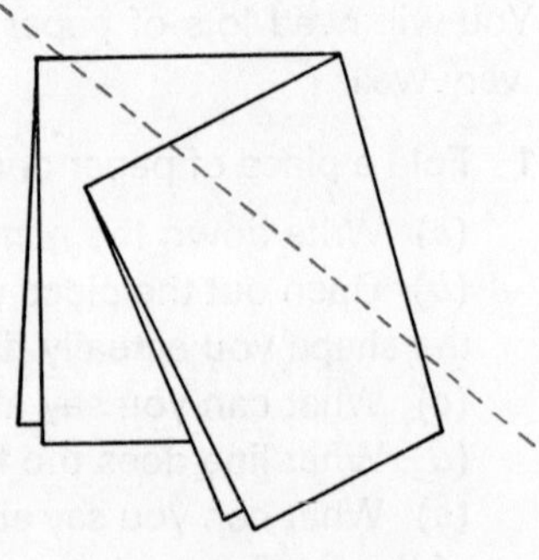

Fig. 8

5 Fold your paper in four again but this time with the second fold oblique (see Figure 8). Cut the corner as shown.

(*a*) How many lines of symmetry has the figure you have made?
(*b*) Are all the folds lines of symmetry? If not, what are they?
(*c*) Could you cut the corner so as to make a triangle?

If you find part (*c*) rather hard, try to make the figure shown in Figure 9 first, then try again to make a triangle.

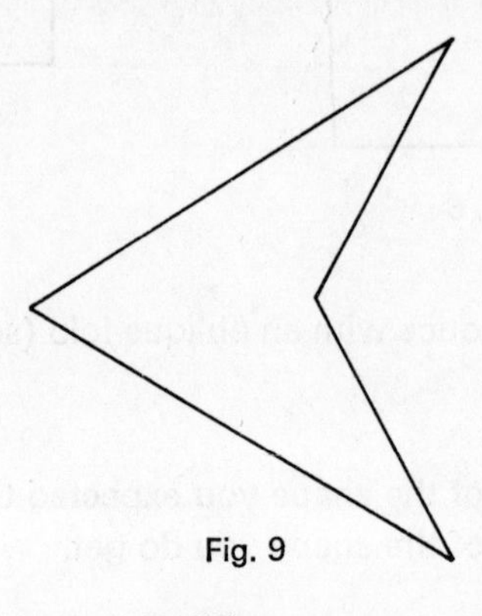

Fig. 9

3 (*a*) Triangle. (*b*) Triangle.

The fold line is the bisector of the angle through which it passes.

4 (*a*) Pupils might suggest a square, diamond or parallelogram.
(*b*) Rhombus.
(*c*) Diagonals—also lines of symmetry.
(*d*) Cut at 45° to fold lines. No.

5 (*a*) One.
(*b*) No. Angle bisectors.
(*c*) The cut should be at 90° to the inner folded edge.

6 (*a*) Usually 6 pointed—could be 8.
(*b*) 6 points.
(*c*) Usually 8 points.

The patterns constructed for Question 6 could be kept as they will be useful later for discussions on rotational symmetry.

2.4 Folding without cutting

Many pupils are surprised by the idea that a line segment could have its own line of symmetry. The perpendicular bisector of the line is easily shown by folding.

From any point on the bisector, lines may be drawn to the ends of the line segment *AB* thus giving the isosceles triangle as discovered in Exercise C, Question 1.

The discovery of the circumcentre is easily made, and, since no special measurements are given for the original triangle, many different examples will be obtained from the pupils.

Owing to the possibility of this centre being external to the triangle it is better not to use cut-out triangles for this exercise.

From considerations of symmetry the pupils will see that the circumcentre is equidistant from all three vertices. Should they see the possibility, they may draw in the circumcircle but there is no need to introduce the idea to those who have not discovered it for themselves.

6 By folding the paper more and more times you can make many different shapes. Try to make the following:

(*a*) a star (all the folds should go through one point which should be cut out);
(*b*) a snowflake (this is like a star with six jagged points; cut small pieces out of the fold lines before you unfold it);
(*c*) a lacy mat.

2.4 Folding without cutting

We have found that we can make four right-angles at a point by folding the paper at that point. In the answers to Exercise C you found that the fold lines halved the angles of your triangles or quadrilaterals (four-sided figures). It is possible to do much more halving by this method and some very interesting results can be found.

When something is divided into two parts in such a way that the two parts are the same, we say that it has been *bisected.*

In Figure 10, AB is a line segment. When we say 'the line AB' this means an endless straight line through the points A and B. But when we are only considering points A and B and the part between them, we refer to 'the line segment AB'.

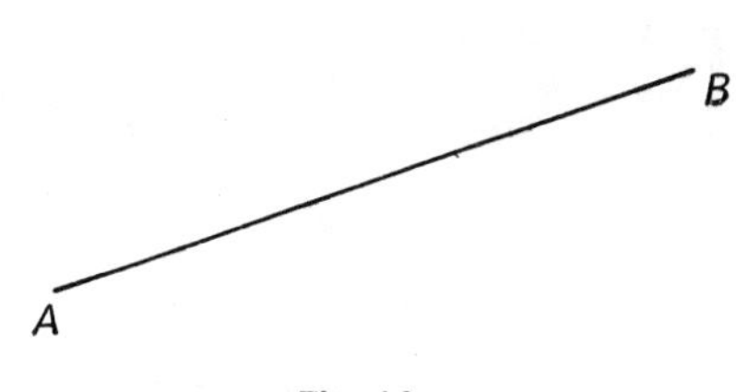

Fig. 10

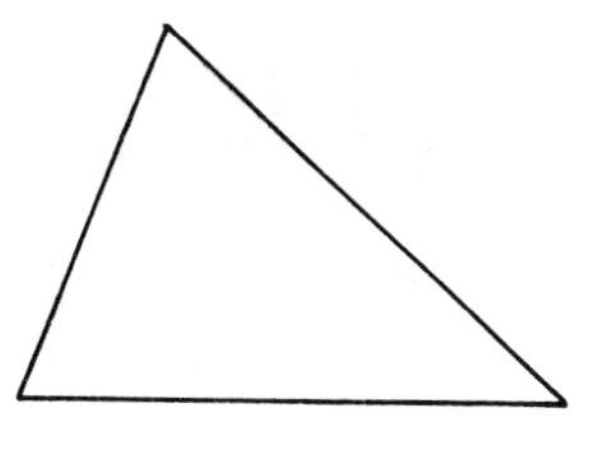

Fig. 11

Has this line segment a line of symmetry? Just looking at it, this does not seem likely and yet it is possible to divide a straight line into two exactly equal parts. Draw a line segment on tracing paper and fold to find the centre. The fold will continue on either side of the centre point. Measure the angles on either side of the fold, between the fold and the line segment AB, with a protractor. Can you think of any other lines you could draw to show that this fold is really a line of symmetry?

The triangle shown in Figure 11 is made from three line segments. Use tracing paper or a triangle cut out of plain paper to find the line of symmetry of each of the three sides of the triangle. Do you notice anything about your result?

Exercise D

(*Results are often more clearly shown if lines of symmetry are drawn in colour*)

1 Draw two triangles which are very different in shape (like those in Figure 12), and find the lines of symmetry for each of the sides of each triangle. Do these lines of symmetry intersect at one point? Is this point inside or outside the triangle?

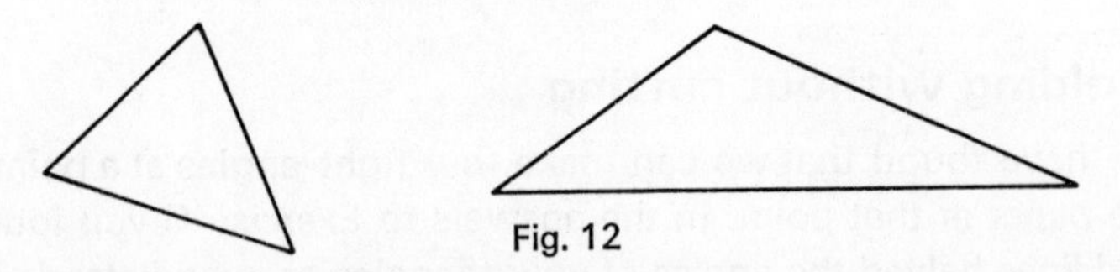

Fig. 12

2 Draw a circle by drawing round a circular object. Find the centre by folding the paper.

3 Copy this set of letters of the alphabet. Try to find lines of symmetry by folding. Remember that not all lines of symmetry are vertical or horizontal and that some shapes may have more than one line of symmetry.

A B C D E F G H I

J K L M N O P Q R

S T U V W X Y Z

4 Find as many lines of symmetry as you can for each of the following figures. Copy them and sketch in the lines.

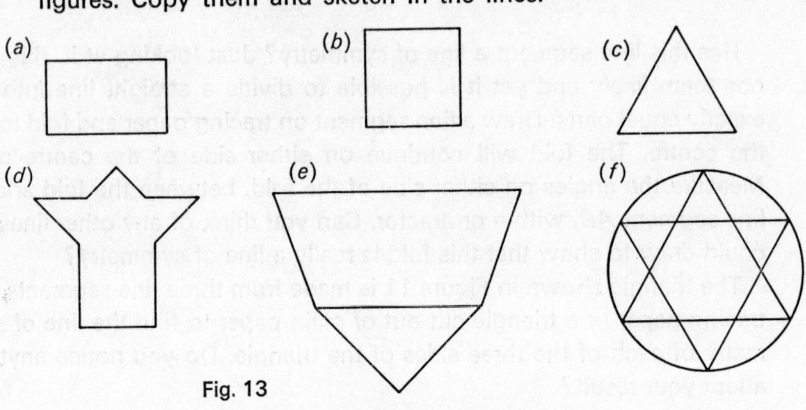

Fig. 13

Exercise D

1 It may be found helpful to warn the pupils to leave plenty of space around their diagrams so that there will be room for the circumcentre to to show clearly on the page.

3 Vertical lines

A H I M T U V W X Y

Horizontal lines

B C D E H I X

The Q has one diagonal line of symmetry. Notice that the X, as drawn in the text, loses its diagonal symmetry because of the thickness of the lines.

O has innumerable lines of symmetry.

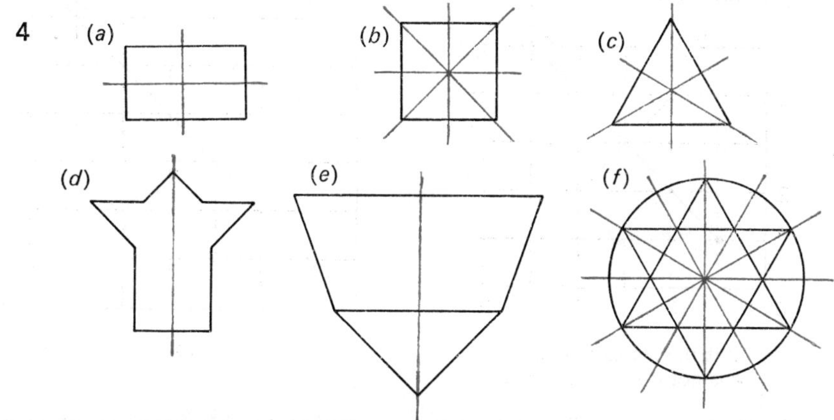

2.5 More about bisecting

Practice is needed at this stage, in finding angle bisectors by folding. From the pupil's standpoint it is easier to fold

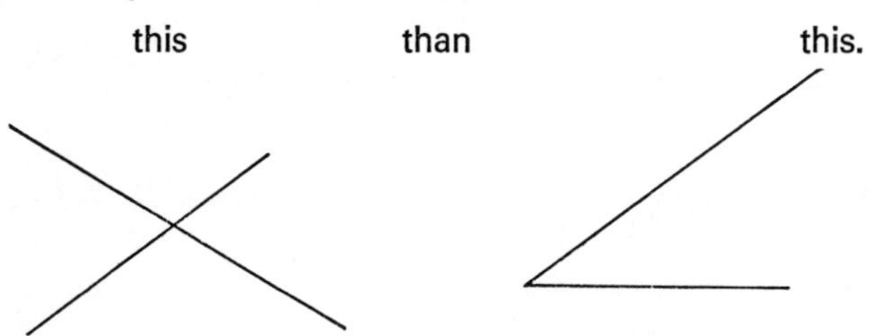

A useful guide to this sort of work is given in *Geometry without Instruments* by K. Lewis.

Exercise E

1

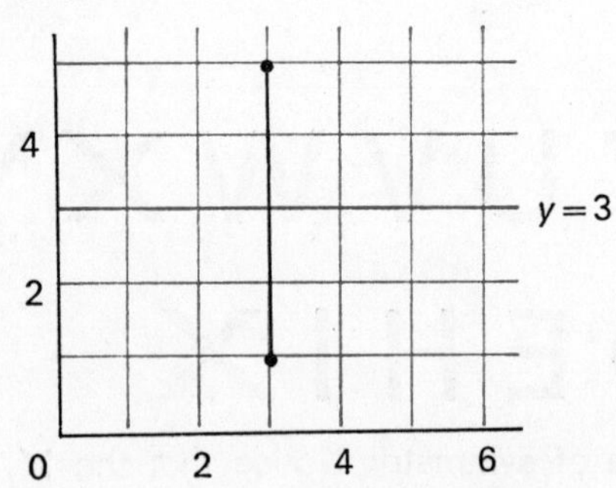

2

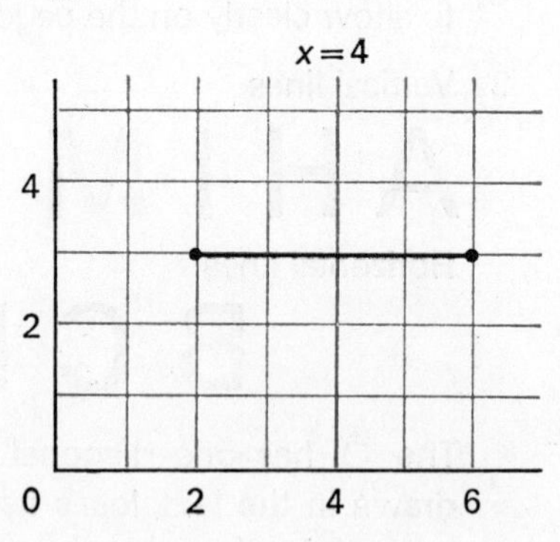

3

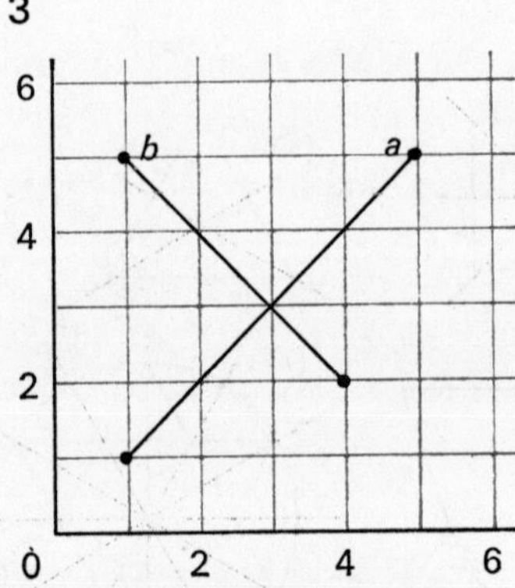

b is the mediator of *a*
a is not the mediator of *b*

4

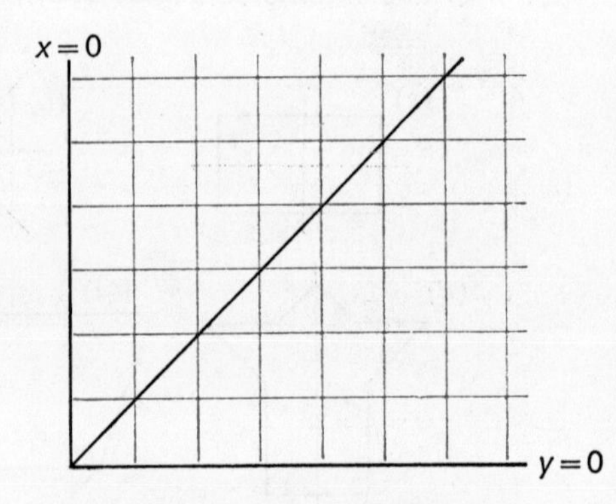

(0, 0), (1, 1), (2, 2), (3, 3), etc.

5 Careful work will produce the result that the bisectors are always at right-angles to each other.

2.5 More about bisecting

In Section 2.4 we learnt about bisecting line segments. The fold line, or line of symmetry, is called the *mediator* of the line segment. The line which cuts an angle into two exactly equal parts is called an *angle bisector.*

For this next exercise you will need pinboard or squared paper.

Exercise E

1 Join the points (3, 1) and (3, 5). Draw in the mediator of this line segment. What is its equation?

2 Draw the line segment joining the points (6, 3) and (2, 3). Draw in the mediator and then write down its equation.

3 Draw the line segment joining the points (1, 1) and (5, 5) and the line segment joining (4, 2) and (1, 5). Is each line the mediator of the other?

4 Draw the two lines $x = 0$ and $y = 0$. Now draw in the bisector of the angle between them. Name three points on this line.

5 Draw two lines that cross (the whole lines in Figure 14), then fold to find both the angle bisectors. (These are shown as broken lines.) Measure the angle between these bisectors. Draw some more pairs of lines and fold to find their angle bisectors. Can you state a fact which you think will apply to all sets of bisectors of this sort?

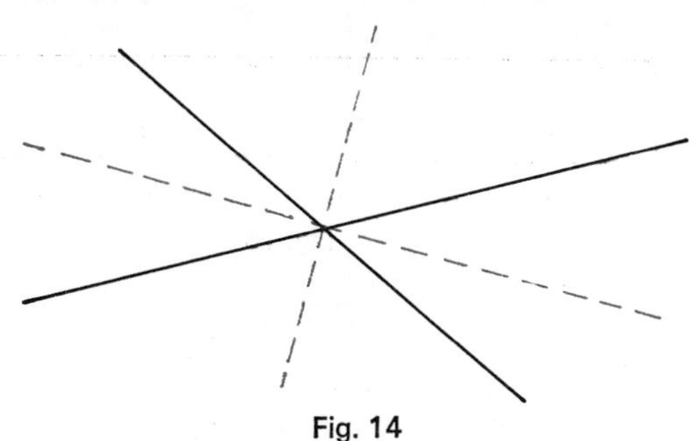

Fig. 14

6 The four corners of a quadrilateral are at the points given by (1, 1), (2, 0), (2, 2), (4, 1). Join these points to form the quadrilateral (it is known as a kite) then draw in any lines of symmetry and state their equations.

7 Copy Figure 15 on page 78. The point (4, 5) is a point at the corner of a figure. The lines $x = 2$ and $y = 4$ are the lines of symmetry of this figure. Find the other corners and then join the points to complete the figure. What is the name of the figure you have drawn?

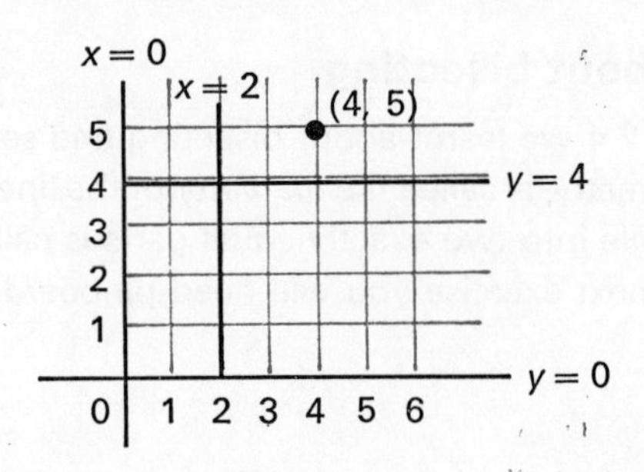

Fig. 15

8 Copy Figure 16. One circle has a radius of 1 centimetre and its centre is the point (2, 1); the centre of the other circle is (4, 2) and its radius 2 centimetres. Draw in the line of symmetry of the figure.

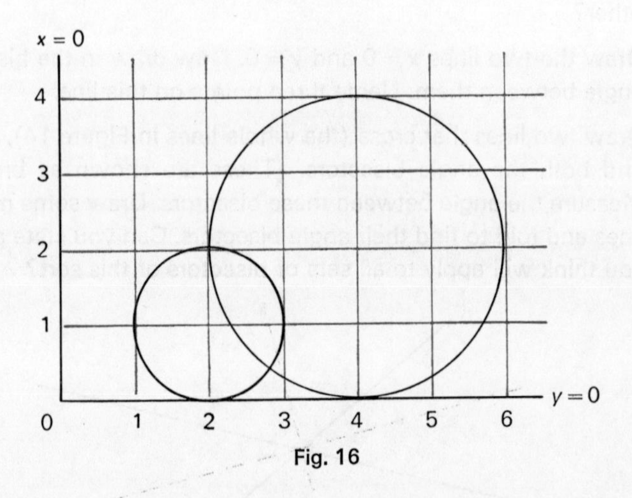

Fig. 16

3. ROTATIONAL SYMMETRY

Trace Figure 17 on two separate pieces of tracing paper. Place one exactly over the other and put a pin through the centres of the figures.

Rotate the top figure until it again covers the lower figure. What fraction of a whole turn has been made?

Again rotate the top figure until it covers the lower figure. What fraction of a turn has been made this time? Is the top figure now in its original position?

6

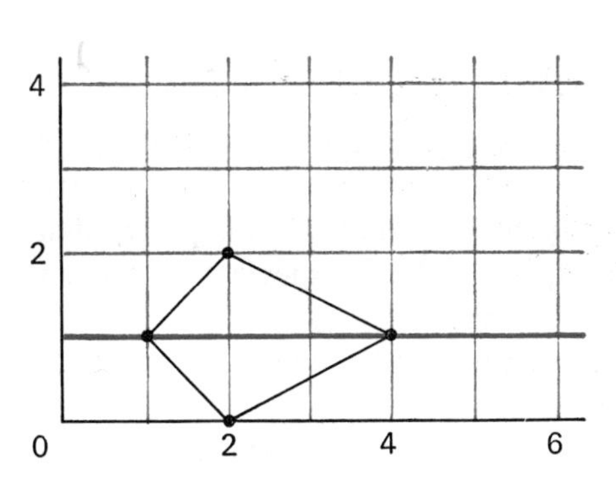

$y = 1$ is the only line of symmetry

7

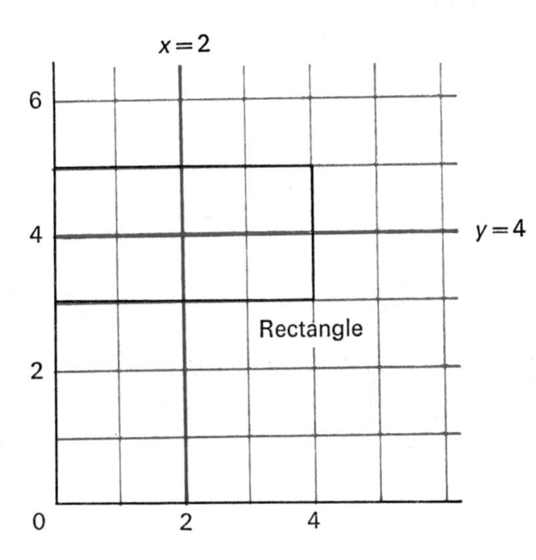

8

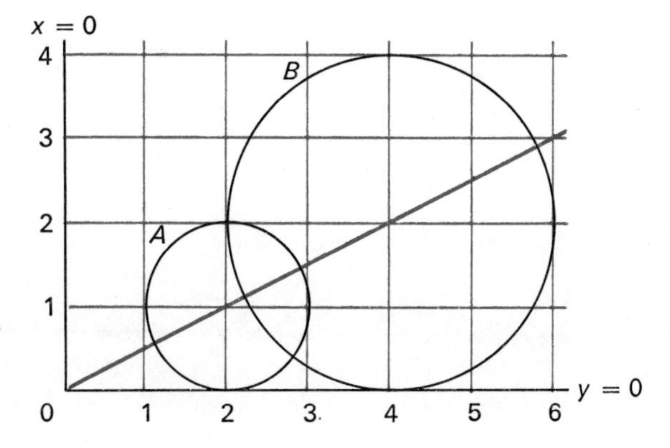

3. ROTATIONAL SYMMETRY

The idea of rotational symmetry holds a strong appeal for most pupils. They feel that certain patterns must be symmetrical and yet there are no lines of symmetry to be found. The discovery that this feeling can be justified is usually greeted with satisfaction and an enthusiastic investigation of all possibilities.

Figures 17, 18 and 19 are intended for class discussion and the pupils should be encouraged to think out their own examples. Aeroplane propellers, watch or clock faces without specific numeral markings, snowflakes, and the other designs they have already constructed will all provide a useful basis for discussion.

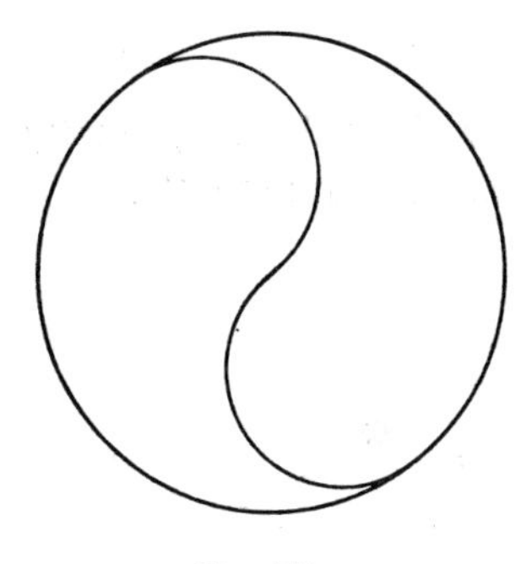

Fig. 17

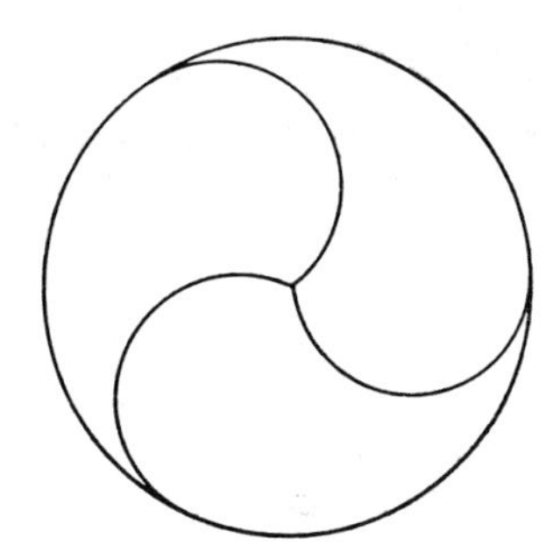

Fig. 18

Make a tracing of Figure 18 and carry out the experiment just described. What fraction of a turn was made on each occasion before the top figure covered the bottom one? How many thirds of a turn had to be made before the top figure was back in its original position?

Neither of these figures has line symmetry but they both have a sort of regularity discovered by rotation. They are said to have *rotational symmetry*. The point about which these figures have to be rotated to discover this sort of regularity is called the *centre of rotation*. (See page 29.)

Because Figure 17 made two moves before it was back in its original position, it is said to have rotational symmetry of order 2. Figure 18 has rotational symmetry of order 3.

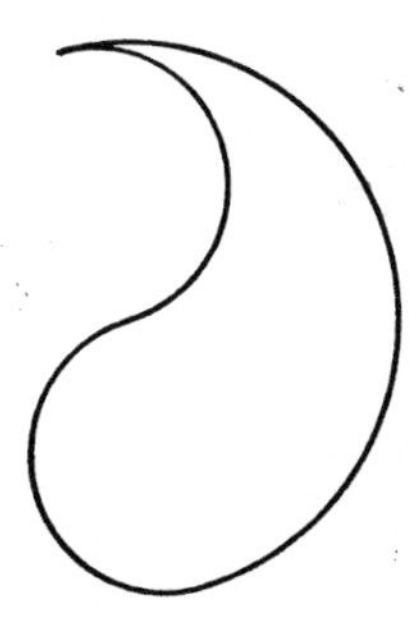

Fig. 19

A figure such as Figure 19 has no rotational symmetry. Copy the figure and choose any point of the plane as a centre of rotation. Could a tracing be rotated about that point so that it falls onto the figure again? Through what part of a turn must it be rotated?

Even figures with no rotational symmetry can be rotated onto themselves by one whole turn. When making tables of the rotational symmetry of different figures, these are given order 1.

Exercise F

1 Use tracing paper to discover the order of rotational symmetry of the following figures about the points marked with red dots.

(*a*)

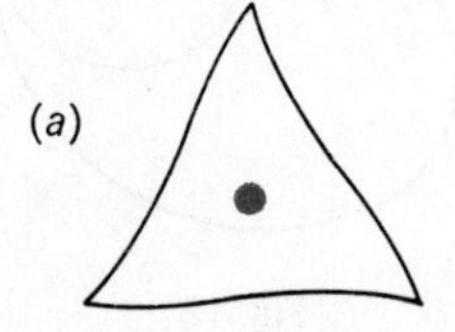

(*b*)

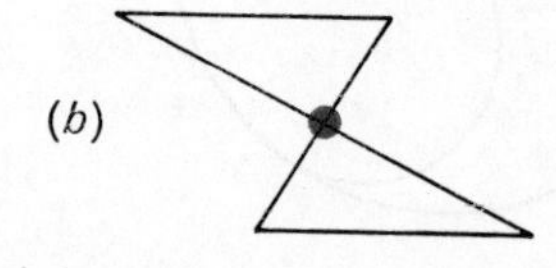

(*c*)

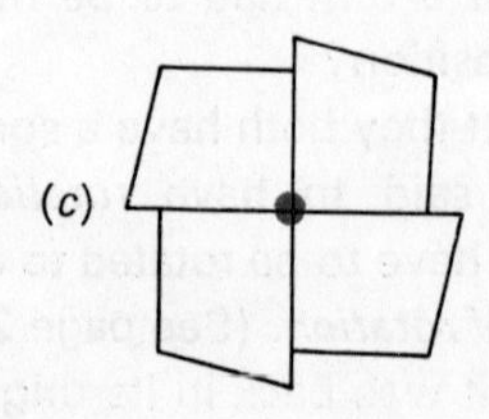

(*d*)

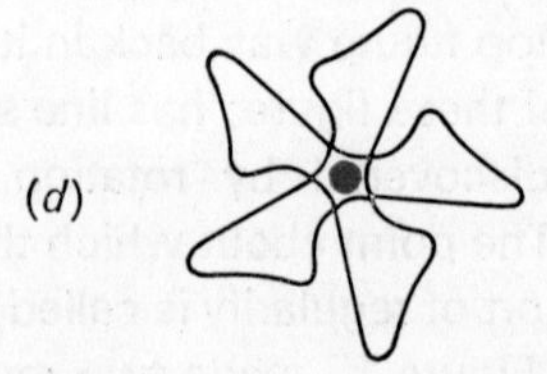

(*e*)

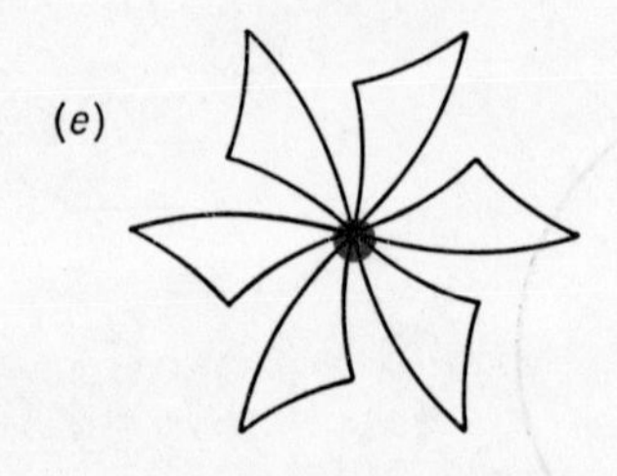

(*f*)

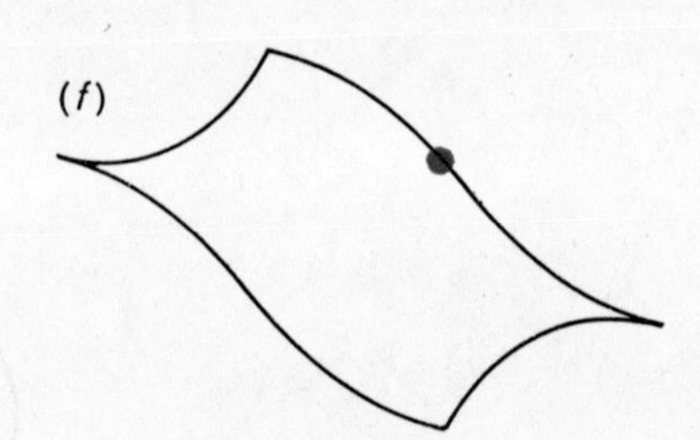

(*g*)

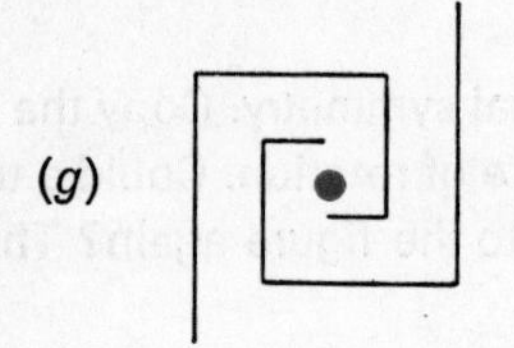

(*h*)

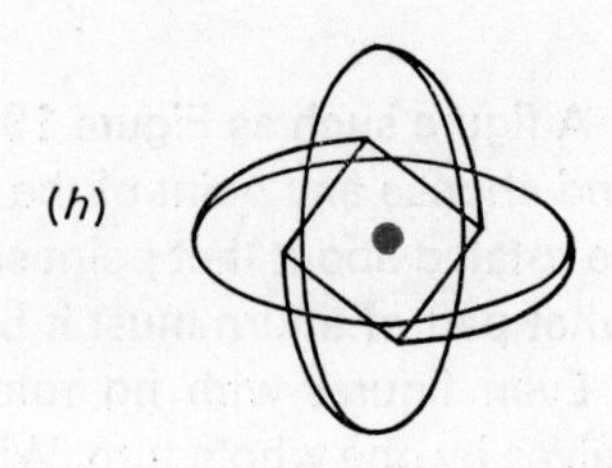

Exercise F

1 (*a*) Order 3; (*b*) order 2;
(*c*) order 4; (*d*) order 5;
(*e*) order 6; (*f*) order 1;
(*g*) order 2; (*h*) order 4.

4. LINE AND ROTATIONAL SYMMETRY

Most pupils will have realized that many patterns they have investigated combine both forms of symmetry, and this section is designed to show some possible combinations and to illustrate the effect of alteration by shading—very useful in designing patterns to given specifications.

Figure 20 (*a*) has 4 lines of symmetry, the two lines given in the figure and the bisectors of the angles between them. The angle between lines is thus 45°.

Figure 20 (*b*) can be altered by shading the section opposite to that already shaded. This is best done by using coloured shading so that the actual shading lines do not cause confusion.

Either of the remaining sections can be shaded to give only 1 line of symmetry but, in order to shade a fourth section to give rotational symmetry of order 2 it is necessary that the two pairs of opposite shadings should match. There will then be two lines of symmetry.

4. LINE AND ROTATIONAL SYMMETRY

Figure 20 (*a*) has rotational symmetry of order 4. But it also has line symmetry. How many lines of symmetry has it? What is the angle between each line of symmetry and the next?

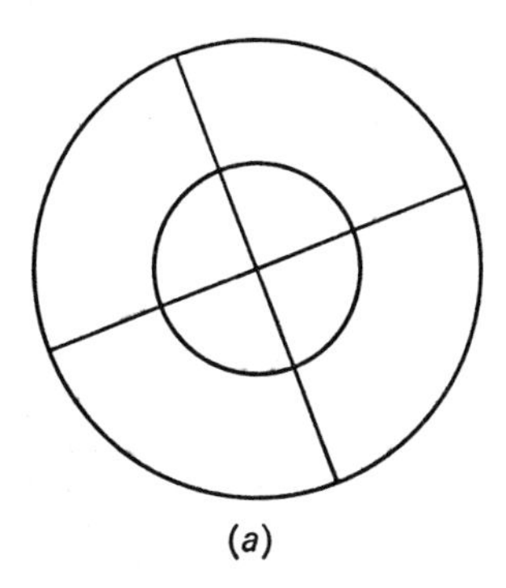

(*a*)

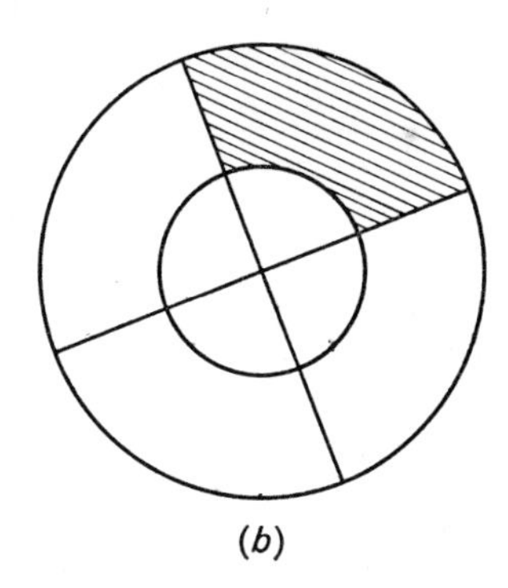

(*b*)

Fig. 20

In Figure 20 (*b*), a section has been shaded. Think of the shading as having changed the figure. This figure has no rotational symmetry and it has only one line of symmetry. Make a copy and shade another section to obtain a new figure with two lines of symmetry and rotational symmetry of order 2. Shade a third section so that the figure again only has one line of symmetry and loses its rotational symmetry. Can you shade a fourth section to give rotational symmetry of order 2? How many lines of symmetry has it now?

Exercise G

1 Trace this pattern. On your tracing draw as many lines of symmetry as you can find. Does this pattern have rotational symmetry about its centre? If so, of what order is it?

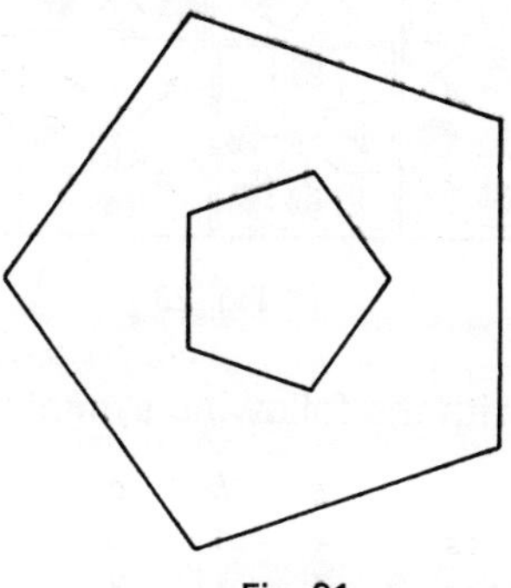

Fig. 21

2 List the symmetries of the following figures. Show your results in a table like this:

Figure	Lines of symmetry	Order of rotational symmetry
a	0	2
b		
c		
d		
e		
f		

(*a*) (*b*) (*c*)

(*d*) (*e*) (*f*)

Fig. 22

3 List the symmetries of the eight diagrams in Figure 23 as in Question 2.

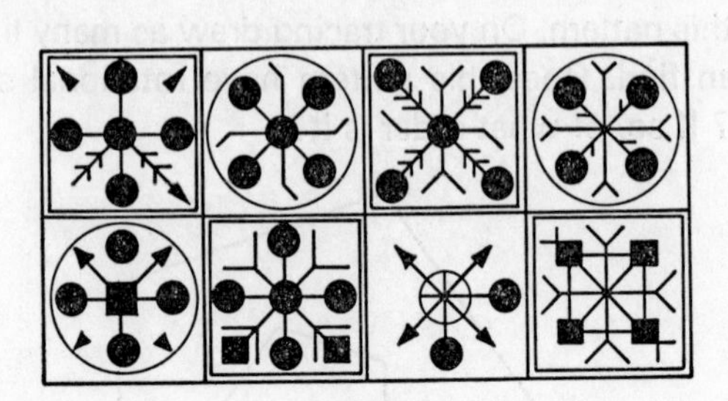

Fig. 23

4 Design patterns with the following symmetries:

	a	*b*	*c*	*d*	*e*	*f*	*g*
Number of lines	4	1	3	2	0	1	6
Order of rotation	4	1	3	2	5	2	6

Exercise G

1 5 lines of symmetry.

Rotational symmetry of order 5.

2

Figure	Lines of symmetry	Order of rotational symmetry
a	0	2
b	1	1
c	0	6
d	2	2
e	0	1
f	8	8

3

Figure	Lines of symmetry	Order of rotational symmetry
1	0	1
2	0	4
3	0	2
4	0	4
5	1	1
6	1	1
7	1	1
8	2	2

(Figures are numbered from top left to bottom right.)

4 *f* is impossible.

Symmetry

5

	Figure	Lines of symmetry	Order of rotational symmetry	Comments
(*a*)	Square	4	4	{2 mediators 2 angle bisectors
(*b*)	Rectangle	2	2	2 mediators
(*c*)	Parallelogram	0	2	
(*d*)	Equilateral triangle	3	3	3 mediator which are also angle bisectors
(*e*)	Isoscles triangle which is not equilateral	1	1	1 mediator which is also an angle bisector

6 (*a*) (i) 1 line (vertical);
(ii) 2 lines (1 vertical, 1 horizontal), order 4;
(iii) 1 line (bisector);
(iv) 1 line (bisector).
(*b*) (i) No lines; (ii) 1 line; (iii) 1 line; (iv) 1 line.

7 It has 2 lines of symmetry and order of rotational symmetry 2.

5 List the symmetries of the following figures. State whether any of the lines of symmetry are mediators or angle bisectors or both:

(*a*) square; (*b*) rectangle (which is not a square);
(*c*) parallelogram (which is not a rectangle);
(*d*) equilateral triangle;
(*e*) isosceles triangle which is not equilateral.

6 Here are some of the patterns seen during a performance by a Formation Dancing Team. Give the symmetries of these patterns thinking of

(*a*) couples as a unit (one black spot, one red spot);
(*b*) each person separately.

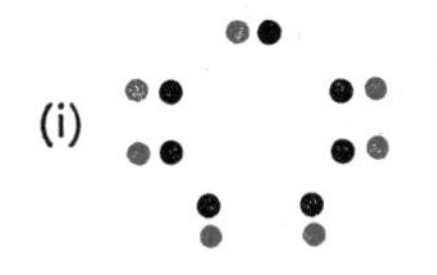

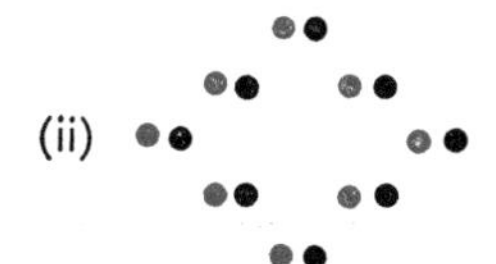

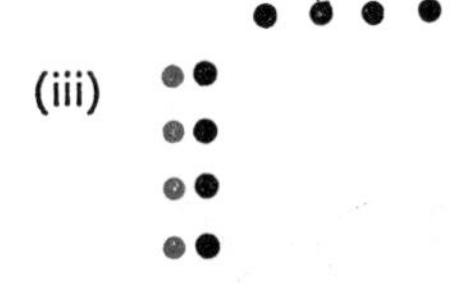

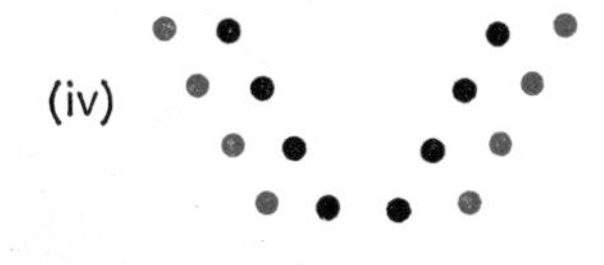

7 Here is a design for a piece of costume jewellery. Discuss its symmetries.

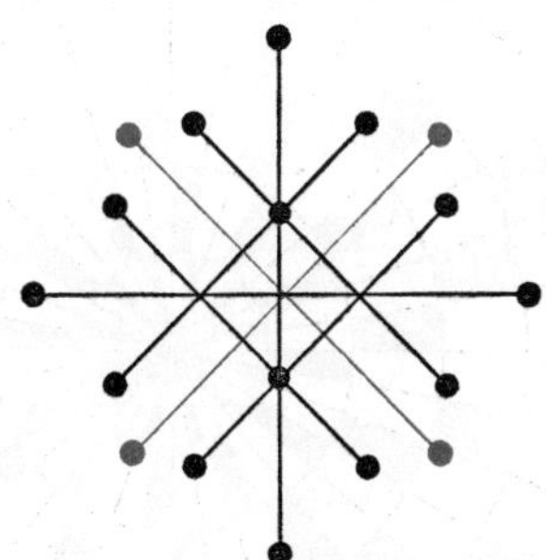

8 Design a piece of costume jewellery to be used as a brooch or a pendant.

9 Design an ornamental star for the top of a Christmas tree.

Interlude

MAKING PATTERNS

Equipment: plain paper; protractor; compasses; ruler. Mark a point near the centre of your paper. Draw straight lines which make angles of 10° with each other at this point. Here are the first five:

Fig. 1

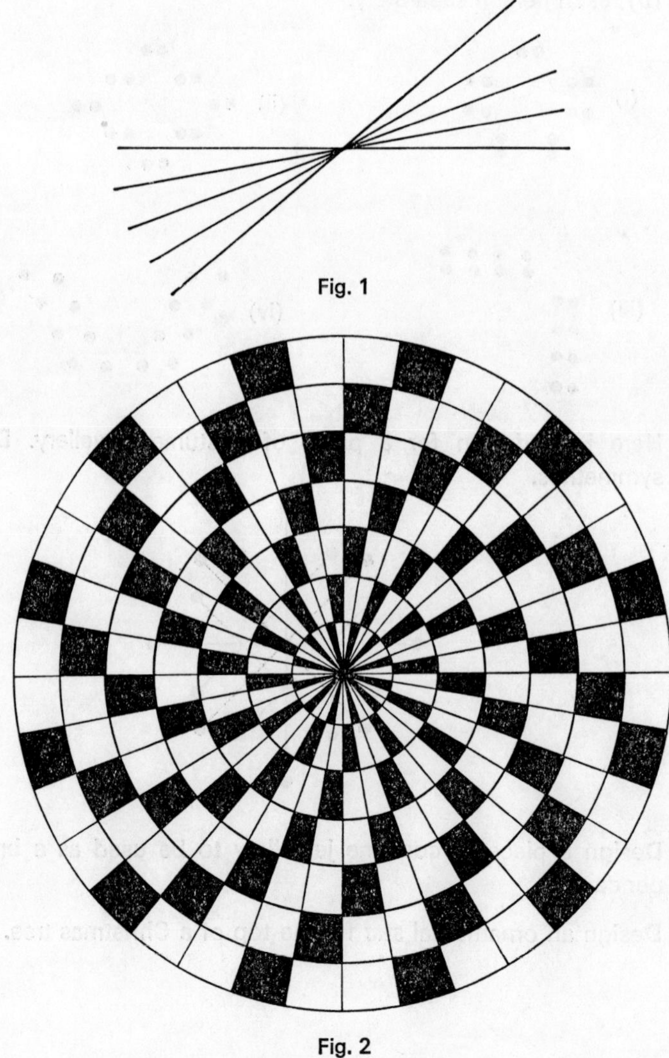

Fig. 2

Using this point as centre, draw circles of radii 1 cm, 2 cm, 3 cm, 4 cm, 5 cm, etc.

Colour the drawing to make many different patterns. Two shown here have a 'spiral' effect (see Figures 2 and 3). Design some 'spiral' patterns of your own. Figure 4 shows another effect that can be achieved.

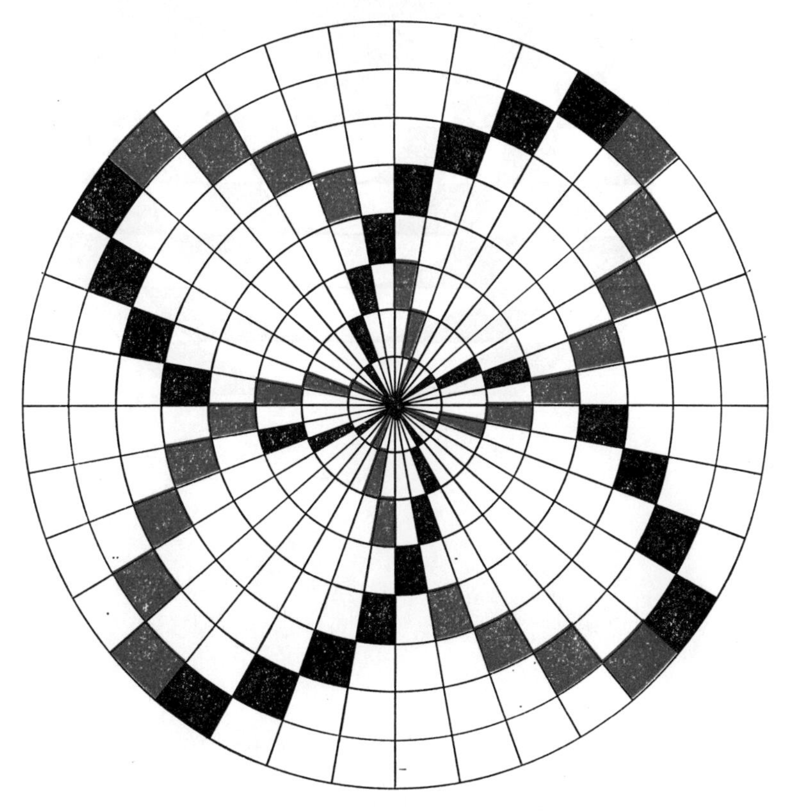

Fig. 3

Fig. 4

Interlude

The principal purpose of this interlude is to draw attention to the enjoyment that may be derived from making patterns, an enjoyment that is often heightened by the use of shading and colour. A subsidiary purpose is to give practice in the handling of compasses, for they are required in constructing regular polygons in a later chapter.

The patterns will almost inevitably have symmetries and where this is the case it is worth trying to describe them. Notice that the symmetries obtained in the use of colour will not necessarily coincide with the symmetries of the figure.

There is no reason why fairly symmetrical polygons should not be formed by the addition of straight lines to the basic construction.

T

Revision exercises

Quick quiz, no. 1

1 132.

2 54°.

3 9, 14, 72.

4 1400_{five}.

5 Line symmetry about the line up the page.

6 (1, 0).

Quick quiz, no. 2

1 15°, 69°, 72°, 89°.

2 13, 26, 39, 52.

3 1041_{five}.

4 2.

5 2.

6 81, 49, 25, 16.

Revision exercises

Quick quiz, no. 1

1 $11 \times 12 = ?$

2 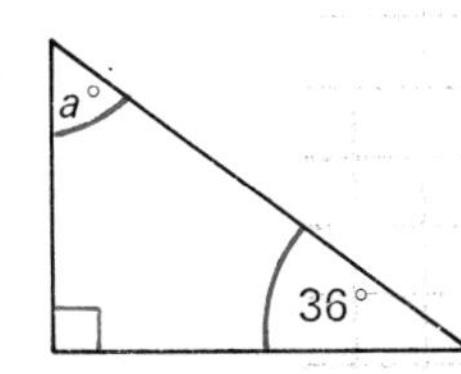

Find a.

3 Which of these are rectangle numbers? 14, 72, 1, 11, 2, 9, 5.

4 $100_{five} \times 14_{five} = ?$

5 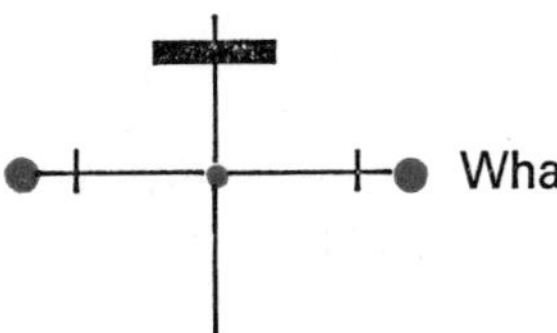

What symmetries has this figure?

6 Three corners of a square have coordinates (0, 0), (0, 1) and (1, 1) respectively. Give the coordinates of the fourth corner.

Quick quiz, no. 2

1 15°, 100°, 340°, 72°, 69°, 89°. Which of these angles are acute?

2 Write down the first four multiples of 13.

3 $140_{five} + 401_{five} = ?$

4 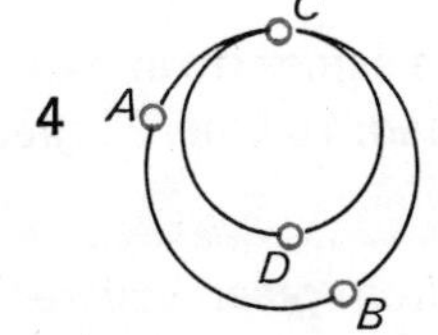

Through how many turns would you rotate in walking round this path in the order *ABCDCA*?

5 How many lines of symmetry has a rectangle?

6 What are the missing numbers? 121, 100, —, 64, —, 36, —, —.

Exercise A

1 (*a*) Join each point to the next; (1, 0), (3, 5), (5, 0), ($\frac{1}{2}$, 3), ($5\frac{1}{2}$, 3), (1, 0). How many lines of symmetry has the figure?

(*b*) Describe this Christmas tree by its coordinates.

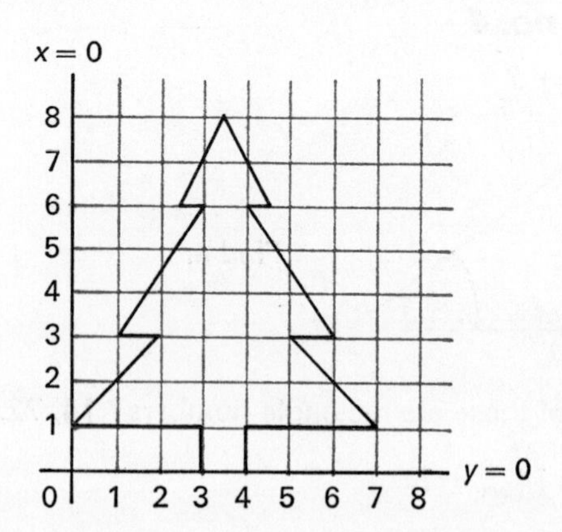

2 (*a*) When you add an odd number and an even number you get an odd number. We can show this on a table thus:

Add	Odd	Even
Odd	—	Odd
Even	—	—

Copy and fill in the rest of the table.

(*b*) Now make a multiplication table for odd and even numbers. One entry is filled in.

Multiply	Odd	Even
Odd	—	—
Even	—	Even

3 Some schoolgirls have knitted 100 squares of the same size in order to make a refugee blanket. 30 of these squares are red, 15 blue, 21 green, 23 yellow and the rest are mauve.

(*a*) What fraction of the blanket will be made from green squares?

(*b*) What fraction of the blanket will be made from mauve squares?

(*c*) What fraction of the blanket will be made from red and blue squares?

Exercise A

1 (*a*) 1.
(*b*) (3, 0), (3, 1), (0, 1), (2, 3), (1, 3), (3, 6), ($2\frac{1}{2}$, 6), ($3\frac{1}{2}$, 8), ($4\frac{1}{2}$, 6), (4, 6), (6, 3), (5, 3), (7, 1), (4, 1), (4, 0).

2 (*a*)

Add	Odd	Even
Odd	Even	Odd
Even	Odd	Even

(*b*)

Multiplication	Odd	Even
Odd	Odd	Even
Even	Even	Even

3 (*a*) $\frac{21}{100}$. (*b*) $\frac{11}{100}$. (*c*) $\frac{45}{100}$.

4 5.

5 (*a*) $a = 28$; $b = 62$.
(*b*) $x = 25$; $y = 118$; $z = 62$.

Exercise B

1 9°; 2·5 kg.

2 (2, 1) and (4, 3). 4. (3, 2).

3 72_{eight}.

4 5 cm.

4 A mysterious race in the Amazonian jungle calculates that

$$21+14=40.$$

How many fingers do you think they are likely to have?

5 Calculate the size of the lettered angles in degrees.

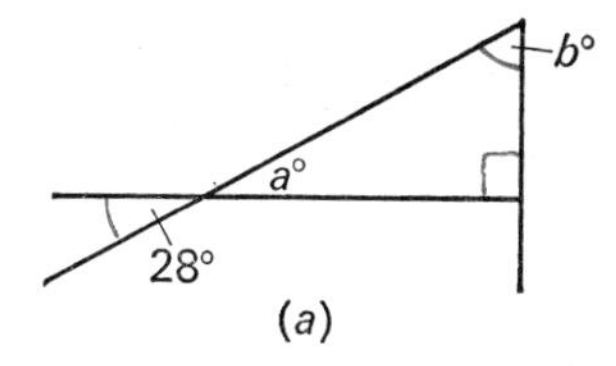

(a)

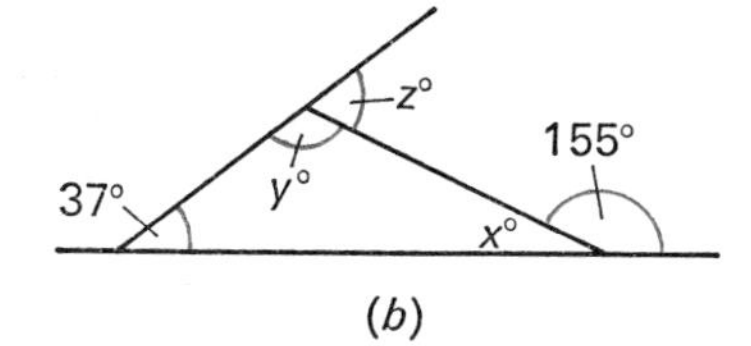

(b)

Exercise B

1 When a turkey weighing 10 kg is placed on a butcher's weighing machine, the pointer on the dial turns through 180°. Through how many degrees will the pointer turn when weighing 0·5 kg of steak?
Another customer buys a joint of lamb and this time the pointer rotates through 45°. How much did this joint weigh?

2 The line $x=3$ is a line of symmetry of a quadrilateral. Two of the vertices are (2, 3) and (4, 1).
Write down the coordinates of the other two vertices.
How many lines of symmetry has the completed figure? What are the coordinates of the point about which the figure has rotational symmetry of order 4?

3 Since spiders have eight legs, we could assume that intelligent spiders would use an arithmetic to the base eight! If flies cost 3_{eight} pence each, bees cost 12_{eight} pence each, find the cost of 6 flies and 4 bees in spiders' pence.

4 If $AC=10$ cm, draw the following figure accurately. Measure CD.

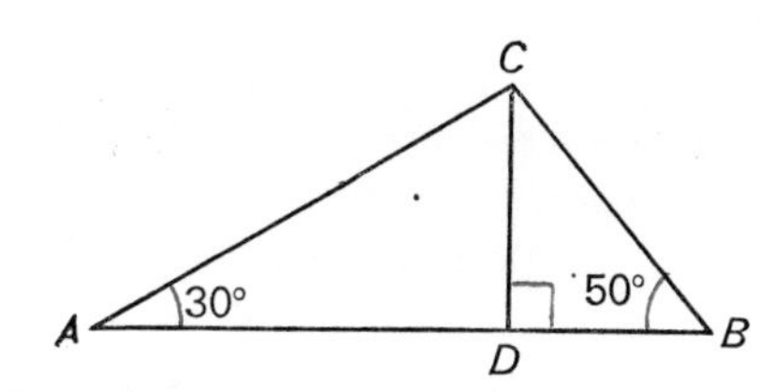

5 How do you get to the centre of this maze? Give your path by stating the coordinates of the squares where you turn.

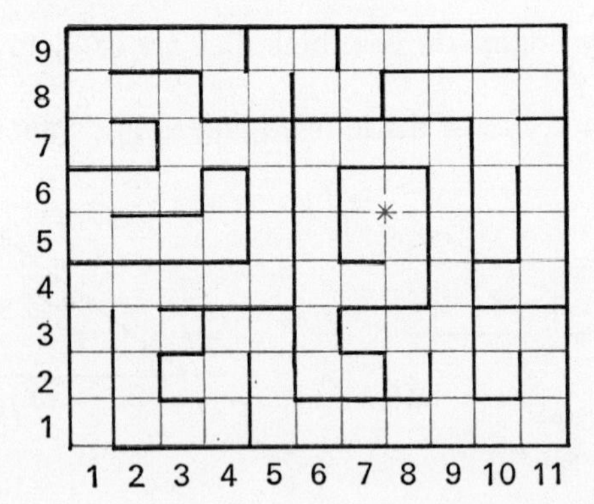

5 (1, 4), (2, 4), (2, 1), (4, 1), (4, 3), (5, 3), (5, 1), (9, 1), (9, 7), (6, 7), (6, 4), (8, 4).

T

6. A quick look at fractions

The purpose of the fraction chapters in this book and the next is to remind the pupils of the meaning of fractions; it is not concerned with complicated techniques for combining them. Sometimes pupils become quite expert at the manipulation of fractions and yet have little understanding of their meaning or of the operations used in combining them. For example, they know perfectly well that two halves make one whole but can still write

$$\tfrac{1}{2}+\tfrac{1}{2}=\tfrac{2}{4}=\tfrac{1}{2}$$

without spotting the contradiction.

We think that pupils should know how to deal with halves, quarters, tenths and similar simple fractions, but that there is no need at this stage for them to be expert at techniques which allow them to combine, for example, elevenths and twenty-ninths.

Most of the work is visual, as much as possible should be practical and experimental, and it should be restricted to common fractions throughout. The first step is to associate a fraction with an expression for parts of a whole. The next is that of comparing fractions with the same denominator and then of adding and subtracting these fractions.

In more formal terms, the idea we are trying to get across is that the fraction $1/a$ is the unique solution of the equation $ax = 1$. (e.g. if there are 7 parts, then each part, x, is given by $7x = 1$, so $x = \tfrac{1}{7}$.) We also want to make clear that we can write b/a in place of $b \times 1/a$, though strictly speaking the former is the unique solution of $ax = b$. (See Chapter 9 in this book.) So, if $b > c$, the pupils will immediately realize that

$$b\times\frac{1}{a} > c\times\frac{1}{a} \quad \text{so} \quad \frac{b}{a} > \frac{c}{a}$$

is obviously true. Similarly $\dfrac{b}{a}+\dfrac{c}{a}$

can be thought of as $b\times\dfrac{1}{a}+c\times\dfrac{1}{a} = (b+c)\times\dfrac{1}{a}.$

(In practice, the application of the distributive law will seem to be obvious here.)

6. A quick look at fractions

1. PARTS OF A WHOLE

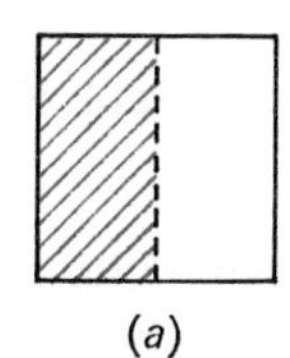

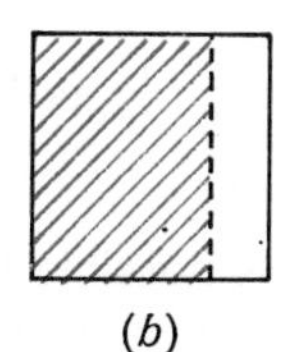

(a) Fig. 1 (b)

Look at these two squares. Each has been split into two parts (shaded and unshaded), but not in the same way. What is the difference between them? In Figure 1 (a) either part is *one-half* of the square, but is this true in Figure 1 (b)?

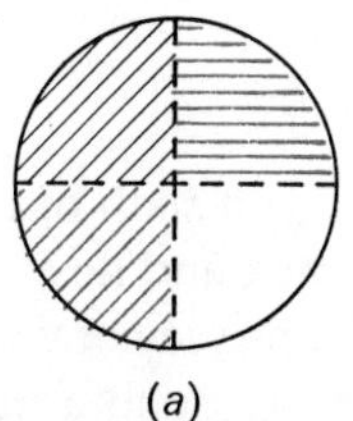

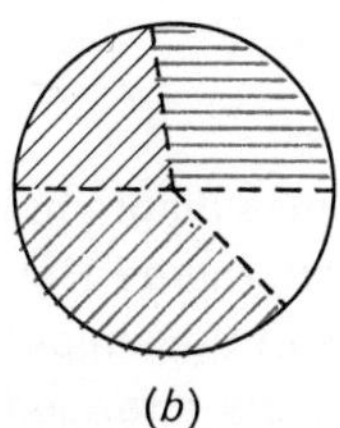

(a) Fig. 2 (b)

Next look at these two circles. Each has been split into four parts. Is there a difference in the way they have been split? In Figure 2 (a) we could choose any part and say that we had *one-quarter* of the circle. Could we do the same in Figure 2 (b)?

Exercise A

1 Which of the following figures have been split into parts of the same size?

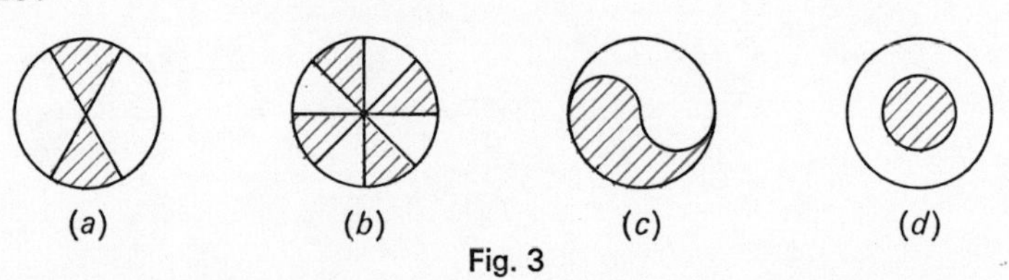

Fig. 3

2 Split each of these figures into the required number of parts of the same shape and size *by eye*.

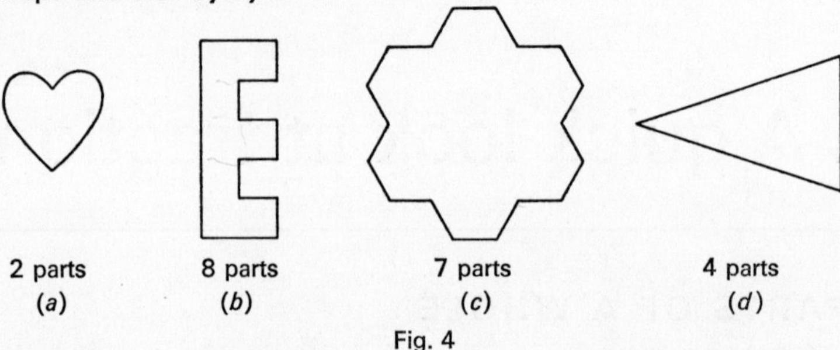

Fig. 4

3 'Peter's got the bigger half', shouted Paul. What is wrong with this remark?

Sometimes it is possible to split a shape into a given number of parts, each of the same size in more than one way. Figure 5 shows three different ways of splitting a disc into 6 equal parts.

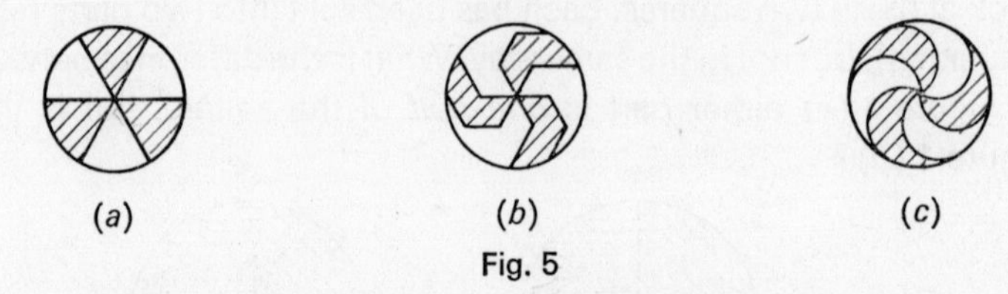

Fig. 5

In any of these diagrams, each part is a sixth of the whole. How many white sixths are there? How many black and how many red sixths?

'Two-sixths of the shape is black.' One way of writing 'two-sixths' is to use two numbers arranged like this: $\frac{2}{6}$. Such a number pair is called a 'fraction'.

The bottom number of a fraction tells us the number of parts into which the shape has been split. (Zero can never be the bottom number of a fraction because it is impossible for one whole to be made from no parts!) The top number indicates how many of these parts we are referring to.

To prove that
$$\frac{b}{a} = b \times \frac{1}{a},$$
the argument is as follows: let $x = 1/a$, then $ax = 1$. Multiply by b so that $a(bx) = b$, then bx is the solution of $ax = b$, namely b/a by definition, and we have
$$b \times \frac{1}{a} = \frac{b}{a}.$$

The arithmetic required should be so familiar that the working really is obvious. We believe that this is no time or place to test the pupil's multiplication tables.

Whenever possible we have avoided technical terms. We refer to top and bottom numbers instead of numerators and denominators because this is how the pupils refer to them. (The comment that numbers as such cannot have positions seems to us irrelevant.)

1. PARTS OF A WHOLE

The first point is that if we are going to use a fraction to express part of a whole, we must divide the whole into *equal* parts.

Exercise A

1 (*b*) and (*c*).

2

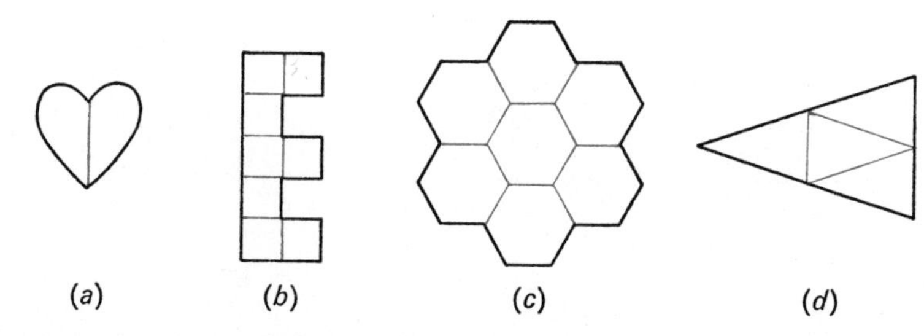

(*a*) (*b*) (*c*) (*d*)

3 We mean by the word 'half', one of two equal parts, and not that a thing is just divided into two.

The point of this section is that a whole can be split into equal parts in a number of different ways. Though it is customary for the equal parts to be of the same shape, it is not necessary.

Exercise B

1 (*a*) $\frac{1}{2}$; (*b*) $\frac{2}{3}$; (*c*) $\frac{5}{12}$; (*d*) $\frac{8}{9}$.

2 (*a*) two-fifths; (*b*) three-quarters;
(*c*) seven-tenths; (*d*) eleven-twelfths.

3 The answers to this question could be given in words or numbers.

(*a*) $\frac{2}{3}$; (*b*) $\frac{1}{4}$; (*c*) $\frac{5}{8}$;
(*d*) $\frac{5}{18}$; (*e*) $\frac{7}{12}$; (*f*) $\frac{7}{15}$.

Example 1

In Figure 5, each red shaded part represents *one-sixth* ($\frac{1}{6}$) of the disc.

Example 2

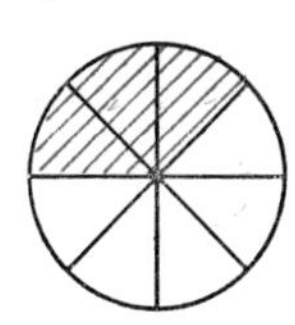

This disc is split into 8 equal parts, with 3 of them shaded. We say that *three-eighths* ($\frac{3}{8}$) of the disc is shaded.

Example 3

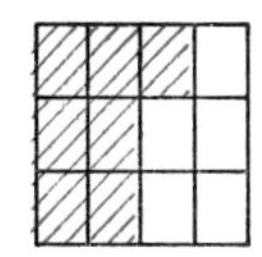

This square is divided into 12 equal parts with 7 of them shaded. We say that *seven-twelfths* ($\frac{7}{12}$) of the square is shaded.

Example 4

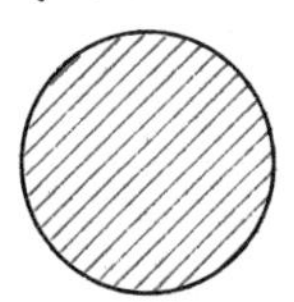

This disc has not been split into parts. The whole is the only part so that 1 whole disc ($\frac{1}{1}$) is shaded.

Exercise B

1 Write each of the following as a fraction:

(*a*) one-half; (*b*) two-thirds;
(*c*) five-twelfths; (*d*) eight-ninths.

2 Write down each fraction in words:

(*a*) $\frac{2}{5}$; (*b*) $\frac{3}{4}$; (*c*) $\frac{7}{10}$; (*d*) $\frac{11}{12}$.

3 What fraction of each figure is shaded?

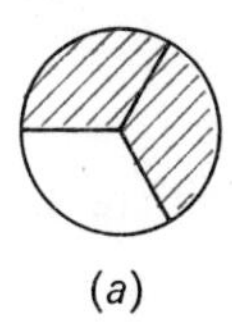

(*a*)

(*b*)

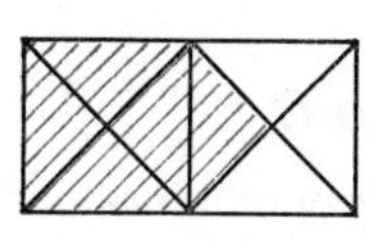

(*c*)

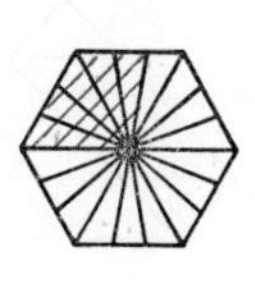

(*d*)

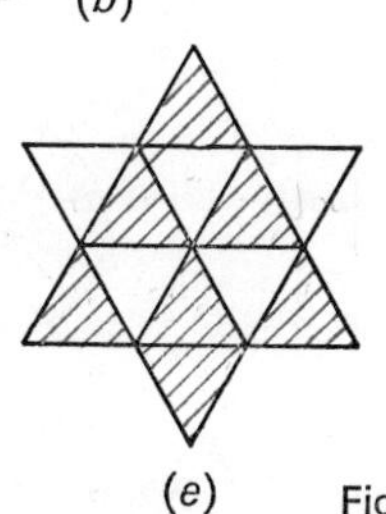

(*e*)

Fig. 6

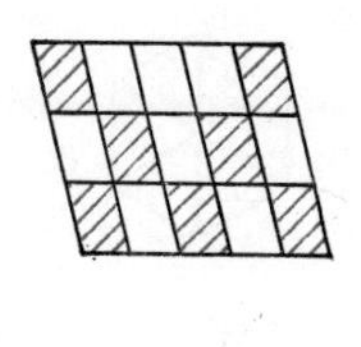

(*f*)

4 Trace each figure and shade the fraction indicated. (Remember that first you must split it into parts of equal shape and size.)

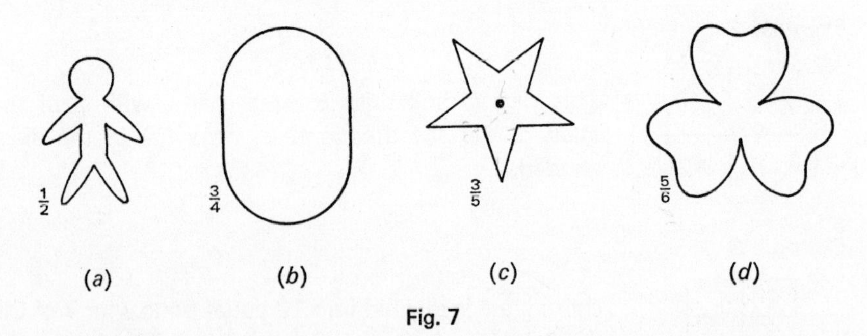

Fig. 7

5 (*a*) What is one-tenth of a centimetre called?
(*b*) What is one-sixtieth of an hour called?
(*c*) What is one-seventh of a week called?
(*d*) What is one-hundredth of a pound (£) called?

The same piece of an object may be described by 'different looking' fractions. Here each fraction gives one-half of the disc.

$\frac{2}{4}$ 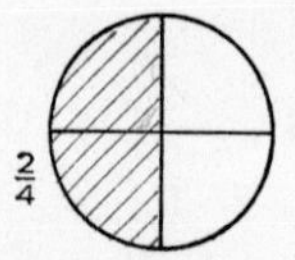 $\frac{4}{8}$ 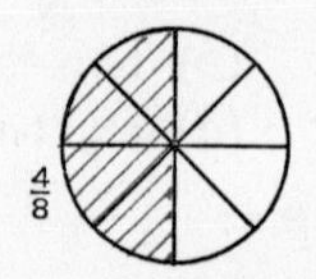 $\frac{3}{6}$ 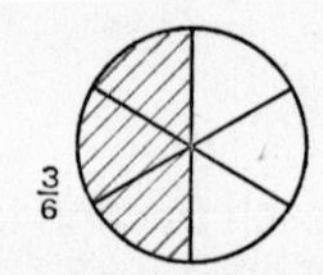 $\frac{6}{12}$ 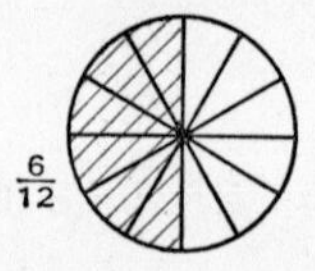

Fig. 8

Exercise C

1 How much of each figure is shaded? Give as many 'different looking' fractions as you can in each case.

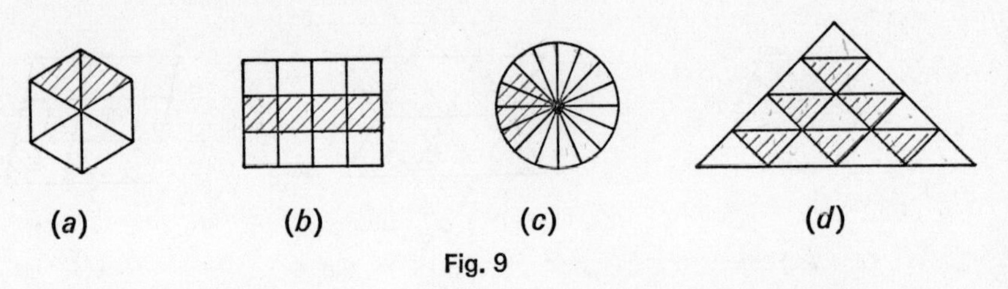

Fig. 9

4

(*a*)

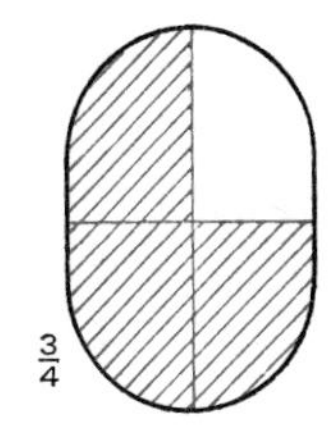

(*b*)

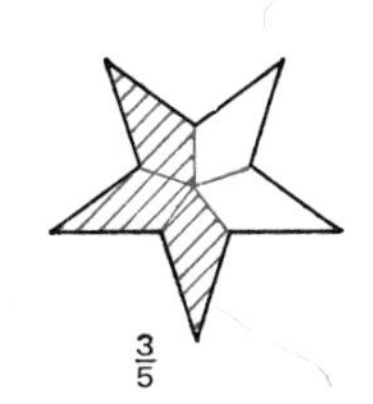

(*c*)

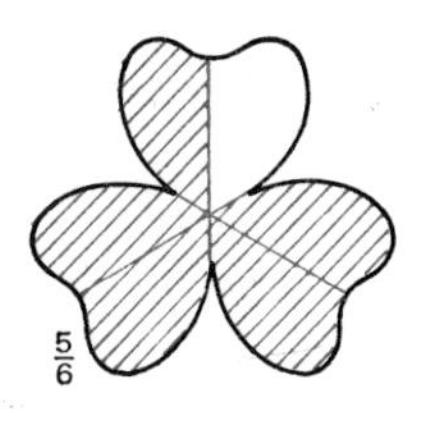

(*d*)

5 (*a*) millimetre;
(*b*) minute;
(*c*) day;
(*d*) (new) penny.

This small section is a reminder that equivalent fractions like $\frac{1}{2}$ or $\frac{2}{4}$ or $\frac{3}{6}$... refer to the same part of a whole. We are not concerned with 'cancelling' or 'simplification'. If pupils mention means of simplification then it might be discussed but there is no need at this stage to emphasize this work.

Exercise C

1 (*a*) $\frac{2}{6}, \frac{1}{3}$; (*b*) $\frac{4}{12}, \frac{2}{6}, \frac{1}{3}$;
(*c*) $\frac{4}{16}, \frac{2}{8}, \frac{1}{4}$; (*d*) $\frac{6}{16}, \frac{3}{8}$.

Any fraction more complicated than the above could be given. For example, in (*a*) the answers might be $\frac{3}{9}, \frac{4}{12}, \frac{5}{15}$, etc.

2. MIXED NUMBERS

In the development of fractions, we do not want to introduce operations or ideas that cannot be covered by some adequate interpretation beyond the fact that 'it works'.

Thus, we shall not be interested in changing $1\frac{2}{3}$ to $\frac{5}{3}$ by 'three times one plus two over three' nor in changing $\frac{4}{6}$ to $\frac{2}{3}$ 'because we can divide the top and bottom of a fraction by the same thing'. We shall only justify these equivalences by visual evidence. In this section, we are developing the idea that it is possible to express whole numbers and mixed numbers in the form of fractions. In this extended sense, $\frac{3}{1}$ and $\frac{3}{2}$ are both to be considered fractions. We cannot change the latter to a mixed number by division because we have not yet associated the operation of division with fractions. We cannot do the cancelling of the last section because we have not yet defined the division of fractions.

In this course, we shall try to develop an understanding of the different number systems, counting numbers, integers, rational numbers, irrationals and real numbers without ever formally describing them. We shall also be developing ideas of structure; for example, relations, functions, operations, identity and inverse. At a later stage these ideas will enable us more satisfactorily to describe the various techniques used in the combination of fractions.

2. MIXED NUMBERS

Example 5

These seven quarters ($\frac{7}{4}$) of a disc together form one and three-quarters ($1\frac{3}{4}$) discs.

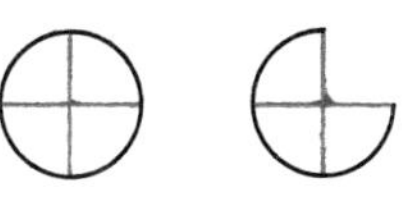

We can write: $\frac{7}{4} = 1\frac{3}{4}$.

Example 6

These eight quarters ($\frac{8}{4}$) of a disc together form two whole (2) discs.

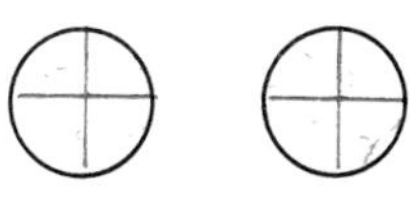

We write: $\frac{8}{4} = 2$.

How do the fractions $\frac{7}{4}$ and $\frac{8}{4}$ differ from the fractions you have met so far in this chapter? Any fraction greater than one is equal *either* to a *'mixed number'* (as in Example 5) *or* a *'whole number'* (as in Example 6).

Example 7

The figure shows a line marked in units and tenths of one unit. We can give the length of the red line as:

(i) a fraction greater than one (that is, $\frac{23}{10}$ of a unit), or
(ii) a mixed number (that is $2\frac{3}{10}$ units).

Exercise D

1 Write down the connection between fractions and mixed numbers illustrated by each of the following diagrams:

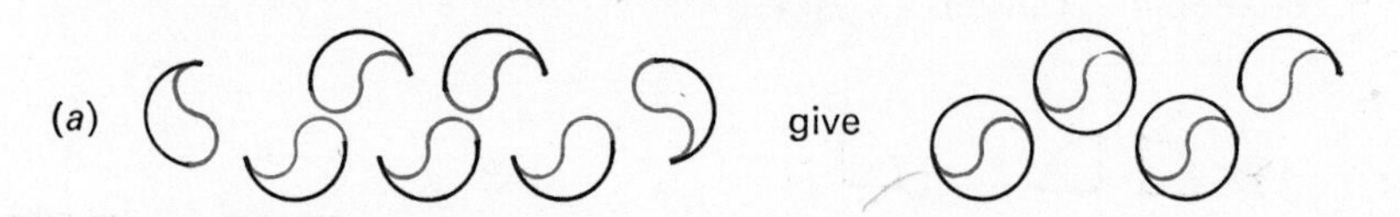

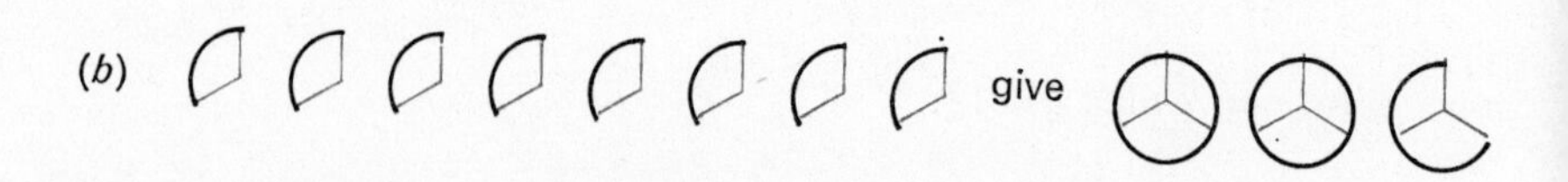

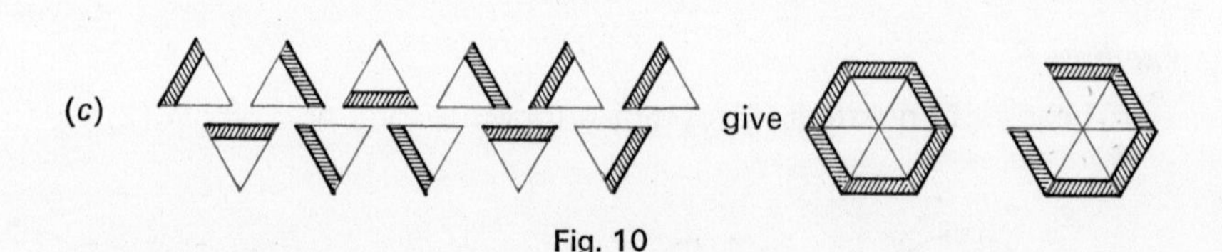

Fig. 10

2 Give each length as a fraction greater than one, and then as a mixed or a whole number.

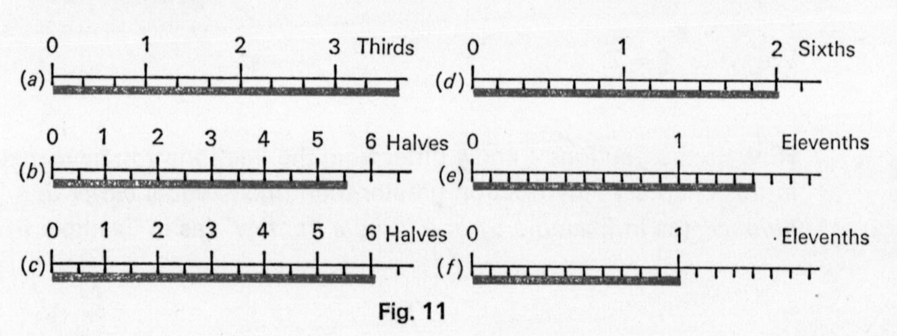

Fig. 11

3 Use the side of an 'old' ruler marked in eighths of an inch to measure the lengths of these lines:

(a) ____________ (c) ______________________________

(b) ____________________ (d) ____________________

Fig. 12

Give the length of each line in two ways (as in Question 2).

Exercise D

1 In this question we hope that the connection between a mixed number and the improper fraction to which it is equivalent will become apparent from the diagrams.

(a) $\frac{7}{2}=3\frac{1}{2}$; (b) $\frac{8}{3}=2\frac{2}{3}$; (c) $\frac{11}{6}=1\frac{5}{6}$.

2 The pupils must first count the total number of thirds, halves, etc. For the second part, they should note the units.

(a) $\frac{11}{3}$, $3\frac{2}{3}$; (b) $\frac{11}{2}$, $5\frac{1}{2}$; (c) $\frac{12}{2}$, 6;
(d) $\frac{12}{6}$, 2; (e) $\frac{15}{11}$, $1\frac{4}{11}$; (f) $\frac{11}{11}$, 1.

3 (a) $\frac{7}{8}$; (b) $1\frac{3}{8}$, $\frac{11}{8}$; (c) 2, $\frac{16}{8}$; (d) $1\frac{4}{8}$ ($1\frac{1}{2}$), $\frac{12}{8}$ ($\frac{3}{2}$).

4 (*a*) Less; (*b*) less; (*c*) equal;
(*d*) greater; (*e*) greater; (*f*) less;
(*g*) greater; (*h*) equal; (*i*) greater.

3. FRACTIONS WITH THE SAME BOTTOM NUMBER

The comparison, addition and subtraction of fractions with the same denominator becomes simply an exercise in the combination of integers once it is realized that these integers refer to numbers of equal parts, just as if they referred to numbers of people or planets. Perhaps pupils should be reminded that these comparisons, additions and subtractions can only be carried out if the parts refer to like objects.

4 Is each fraction less than one, equal to one, or greater than one?

(*a*) $\frac{3}{4}$; (*b*) $\frac{99}{100}$; (*c*) $\frac{100}{100}$; (*d*) $\frac{21}{20}$;
(*e*) $\frac{100}{3}$; (*f*) $\frac{20}{33}$; (*g*) $\frac{15}{5}$; (*h*) $\frac{127}{127}$;
(*i*) $\frac{1001}{1000}$.

3. FRACTIONS WITH THE SAME BOTTOM NUMBER

Fractions with the same bottom number are easy to compare in size. These fractions: $\frac{3}{7}, \frac{5}{7}, \frac{7}{7}, \frac{13}{7}, \frac{22}{7}, \frac{28}{7}$ could all be shown on a number line, marked out in sevenths:

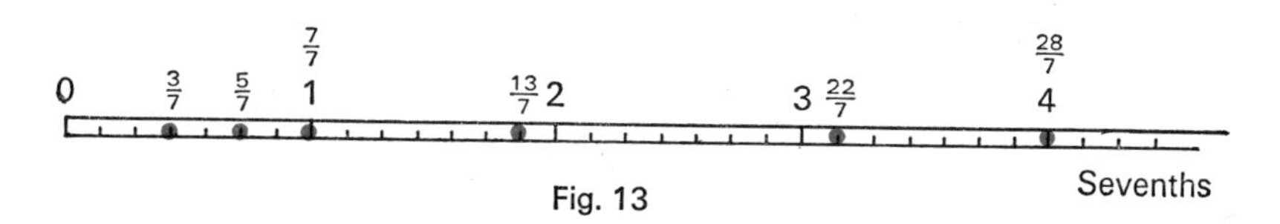

Fig. 13

This shows them in order of size. Numbers to the left are smaller than numbers to the right.

You should now be able to arrange the following fractions in order of size without first marking them on a number line:

$$\frac{2}{9}, \frac{7}{9}, \frac{22}{9}, \frac{4}{9}, \frac{15}{9}.$$

How did you do this?

Fractions with the same bottom number are also easy to add together. This is because we are dealing with parts of the same size. It is not so easy to add together fractions with different bottom numbers. This should be clear if you look at the following diagrams.

Example 8

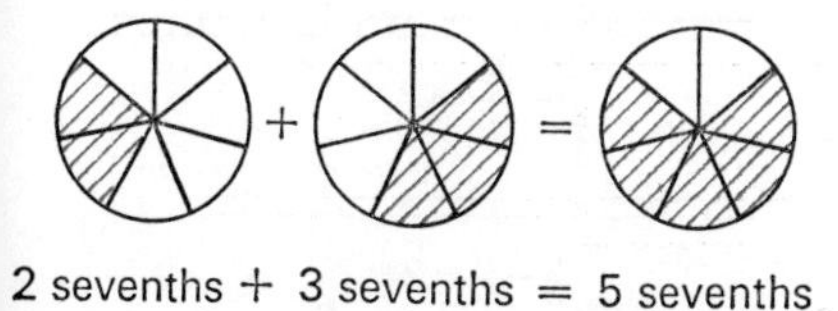

2 sevenths + 3 sevenths = 5 sevenths

Fig. 14 (*a*)

Example 9

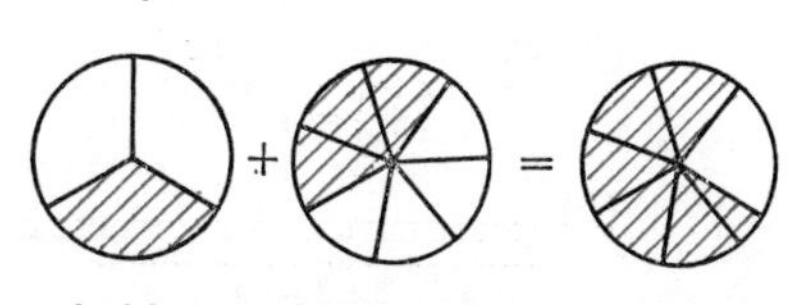

1 third + 3 sevenths = ?

Fig. 14 (*b*)

(You will learn to do the addition of Example 9 later.)

The same thing is true if you want to take one fraction away from another.

Example 10

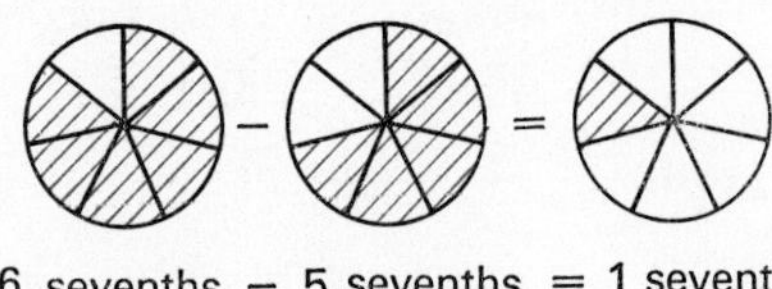

6 sevenths − 5 sevenths = 1 seventh

Fig. 15

Examples 8 and 10 could also have been shown on a number line marked out in sevenths:

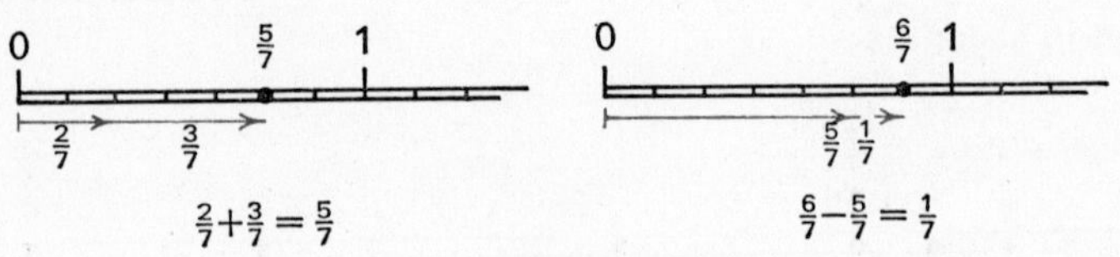

Fig. 16

Exercise E

1 Arrange the following fractions in order of size (smallest first):

$$\frac{13}{13}, \frac{1}{13}, \frac{15}{13}, \frac{6}{13}, \frac{20}{13}, \frac{0}{13}.$$

2 The sign '<' means 'is less than', and the sign '>' means 'is greater than'.

Which of the following statements are true and which are false?

(*a*) 3 quarters > 1 quarter; (*b*) 4 halves < 1 half;

(*c*) 5 tenths < 11 tenths; (*d*) $\frac{5}{7} < \frac{8}{7}$;

(*e*) $\frac{4}{3} < \frac{7}{3}$; (*f*) $\frac{15}{5} > \frac{3}{5}$.

3 Give the addition or subtraction shown on each line.

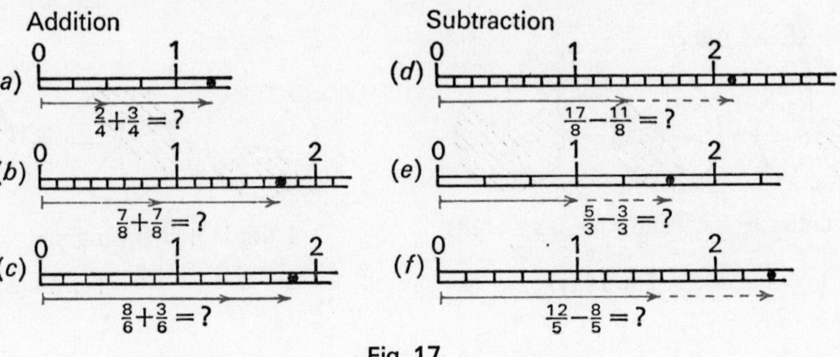

Fig. 17

When can you give your answer as a mixed number?

Exercise E

1 $\frac{0}{13}$, $\frac{1}{13}$, $\frac{6}{13}$, $\frac{13}{13}$, $\frac{15}{13}$, $\frac{20}{13}$.

2 (*a*) True; (*b*) false; (*c*) true;
(*d*) true; (*e*) true; (*f*) true.

3 (*a*) $\frac{5}{4}$; (*b*) $\frac{14}{8}$; (*c*) $\frac{11}{6}$;
(*d*) $\frac{6}{8}$; (*e*) $\frac{2}{3}$; (*f*) $\frac{4}{5}$.
(*a*), (*b*) and (*c*) can be represented as mixed numbers.

4 (*a*) 4 fifths or $\frac{4}{5}$; (*b*) 8 sevenths or $1\frac{1}{7}$;
(*c*) $\frac{8}{100}$; (*d*) $\frac{3}{9}$; (*e*) $3\frac{2}{5}$;
(*f*) 1; (*g*) 0; (*h*) 1;
(*i*) $\frac{2}{7}$.

Exercise F (Miscellaneous)

1 $1\frac{1}{8}$, $1\frac{5}{8}$, $1\frac{7}{8}$, 2, $2\frac{5}{8}$, 3, $3\frac{2}{8}$ ($3\frac{1}{4}$), 4, $4\frac{4}{8}$ ($4\frac{1}{2}$), 5, $5\frac{3}{8}$.

2 (*a*) $\frac{5}{12}$; (*b*) $\frac{4}{12}$ or $\frac{1}{3}$; (*c*) $\frac{3}{12}$ or $\frac{1}{4}$.

3 (*a*) $1=\frac{8}{8}$; (*b*) $2=\frac{16}{8}$; (*c*) $3=\frac{24}{8}$;
(*d*) $3=\frac{6}{2}$; (*e*) $10=\frac{30}{3}$; (*f*) $4=\frac{20}{5}$;
(*g*) $1\frac{1}{2}=\frac{3}{2}$; (*h*) $1\frac{3}{5}=\frac{8}{5}$; (*i*) $2\frac{3}{4}=\frac{11}{4}$.

4 (*a*) $\frac{9}{8}$; (*b*) $\frac{11}{8}$; (*c*) $\frac{15}{8}$;
(*d*) $\frac{19}{8}$; (*e*) $\frac{29}{8}$; (*f*) $\frac{47}{8}$.

5 (*a*) $1\frac{1}{10}$; (*b*) 2; (*c*) $1\frac{7}{10}$;
(*d*) 4; (*e*) $2\frac{3}{10}$; (*f*) $5\frac{5}{10}$ or $5\frac{1}{2}$.

6 $\frac{3}{10}+\frac{3}{10}+\frac{5}{10}=\frac{11}{10}$ which is more than one whole.
Joe, Nancy and Sam together accept $(\frac{3}{13}+\frac{3}{13}+\frac{5}{13})$ of the haul, i.e. $\frac{11}{13}$ of the treasure, leaving $\frac{2}{13}$ for Al.

4 Calculate:

(*a*) 3 fifths+1 fifth; (*b*) 3 sevenths+5 sevenths;
(*c*) 17 hundredths−9 hundredths;
(*d*) $\frac{5}{9}-\frac{2}{9}$; (*e*) $\frac{14}{5}+\frac{3}{5}$; (*f*) $\frac{1}{2}+\frac{1}{2}$;
(*g*) $\frac{2}{3}-\frac{2}{3}$; (*h*) $\frac{1001}{1005}+\frac{4}{1005}$; (*i*) $\frac{11}{7}-\frac{9}{7}$.

Exercise F (Miscellaneous)

1 The points marked in red represent fractions greater than one. Give them as mixed or whole numbers.

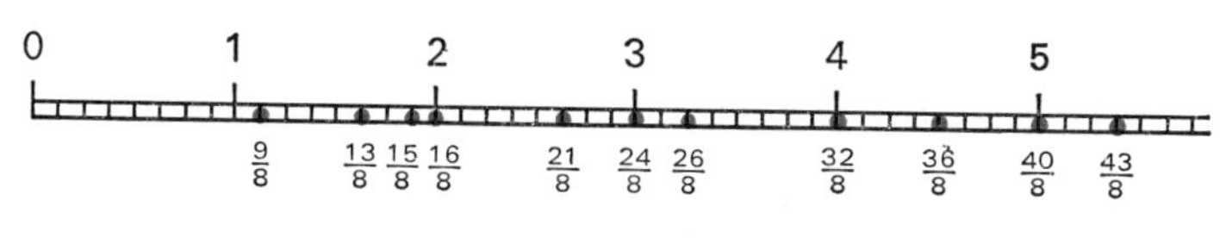

Fig. 18

2 Belinda had an orange consisting of 12 equal pieces.

(*a*) She ate 5 of these pieces. What fraction of the orange had she eaten?
(*b*) Her brother ate 4 of them. How much had he eaten?
(*c*) What fraction of the orange was left?

3 Complete the following:

(*a*) $1=\frac{?}{8}$; (*b*) $2=\frac{?}{8}$; (*c*) $3=\frac{?}{8}$;
(*d*) $3=\frac{?}{2}$; (*e*) $10=\frac{?}{3}$; (*f*) $4=\frac{?}{5}$;
(*g*) $1\frac{1}{2}=\frac{?}{2}$; (*h*) $1\frac{3}{5}=\frac{?}{5}$; (*i*) $2\frac{3}{4}=\frac{?}{4}$.

4 Change each mixed number to a fraction greater than one. (Use the side of an 'old' ruler marked in eighths of an inch to help you if necessary.)

(*a*) $1\frac{1}{8}$; (*b*) $1\frac{3}{8}$; (*c*) $1\frac{7}{8}$;
(*d*) $2\frac{3}{8}$; (*e*) $3\frac{5}{8}$; (*f*) $5\frac{7}{8}$.

5 Change each of these fractions to either a mixed number or a whole number. (Use the side of your ruler marked in tenths of a centimetre to help you if necessary.)

(*a*) $\frac{11}{10}$; (*b*) $\frac{20}{10}$; (*c*) $\frac{17}{10}$;
(*d*) $\frac{40}{10}$; (*e*) $\frac{23}{10}$; (*f*) $\frac{55}{10}$.

6 Al, Joe, Nancy and Sam have just dug up a treasure. Joe and Nancy each want $\frac{3}{10}$ of the money while Sam wants $\frac{5}{10}$ of it. Why is this impossible? Al persuades Joe and Nancy to accept $\frac{3}{13}$ each and Sam to accept $\frac{5}{13}$. How much is left for him? Do you think he was wise?

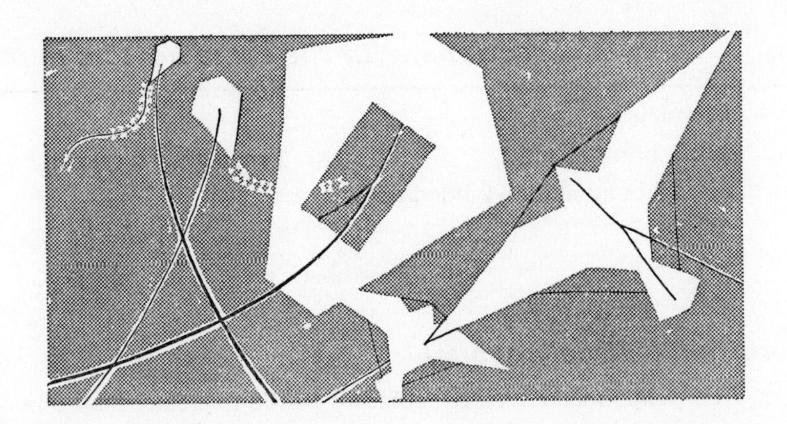

7. Polygons

1. NAMING POLYGONS

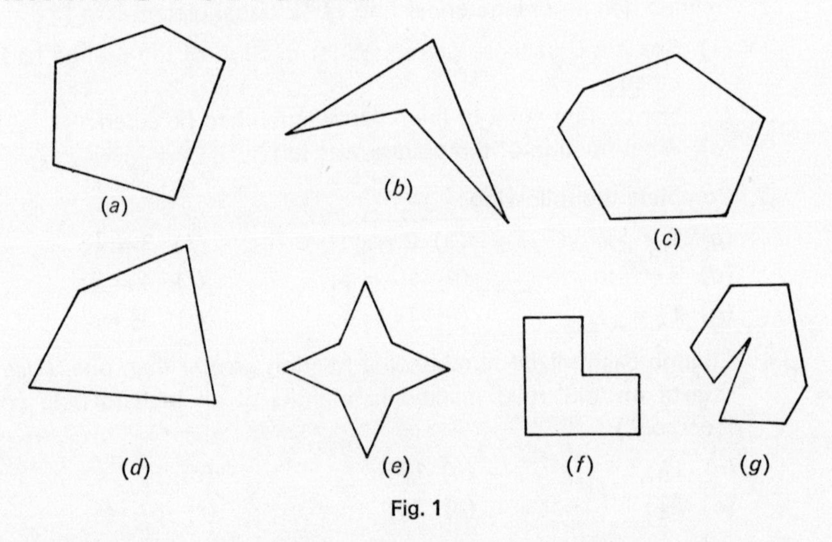

Fig. 1

The plane shapes in Figure 1 are all *polygons*. We use the word 'plane' when the shapes lie in a flat surface, such as the top of a polished table.

The boundary of a polygon is formed from parts (or segments) of straight lines. These line segments are called the *sides* or *edges* of the polygon.

The names of some of the more common polygons are:

Triangle	(3 sides)	Triangle means three angles
Quadrilateral	(4 sides)	Quadrilateral means four sides
Pentagon	(5 sides)	Pentagon means five corners

7. Polygons

Usually, in this chapter, a polygon is thought of as a region together with its boundary of straight line segments. Later work involving the construction of polyhedra and the colouring of tessellations should strengthen this concept. However, in questions concerning frameworks, the word 'polygon' is used to refer to the boundary of a region.

1. NAMING POLYGONS

(*a*) (*a*) Pentagon, (*b*) quadrilateral, (*c*) hexagon, (*d*) quadrilateral, (*e*) octagon, (*f*) hexagon, (*g*) octagon.

(*b*) Pentagon. It is impossible to find two points *A*, *B* of the pentagon so that part of the line segment *AB* lies outside the pentagon.

(*c*) Quadrilateral. Figure A shows two possible positions for the pair of points *A*, *B*.

(*d*) (*a*), (*c*), (*d*).

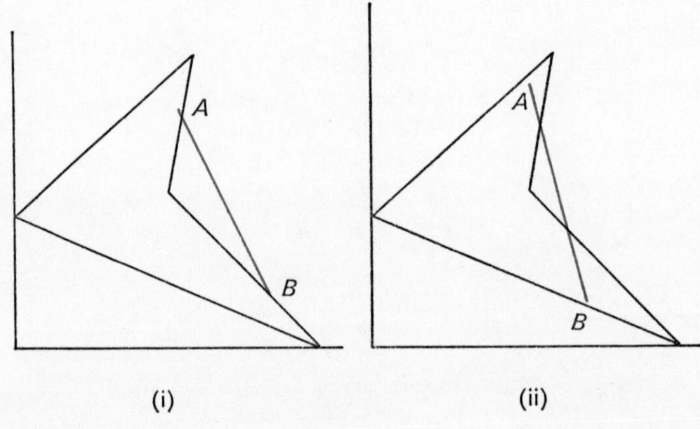

Fig. A

Hexagon (6 sides) Hexagon means six corners
Octagon (8 sides) Octagon means eight corners

(*a*) Name each of the polygons in Figure 1.

(*b*)

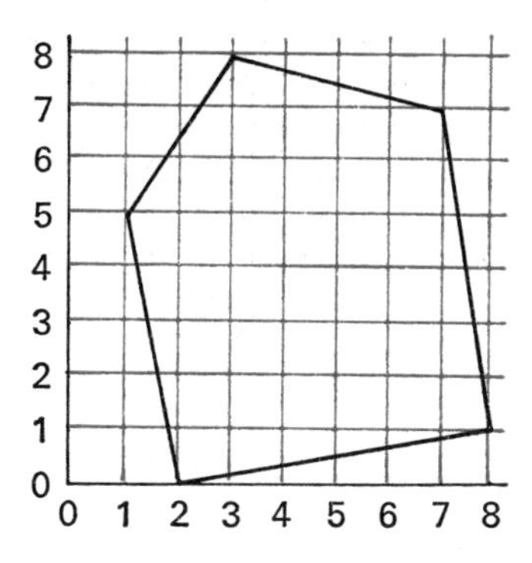

Fig. 2

Copy and name the polygon shown in Figure 2.

A and *B* are *any* two points of the polygon. In Figure 3, *AB* lies entirely within the polygon. Can you find *A* and *B* so that part of the line segment *AB* lies outside the polygon?

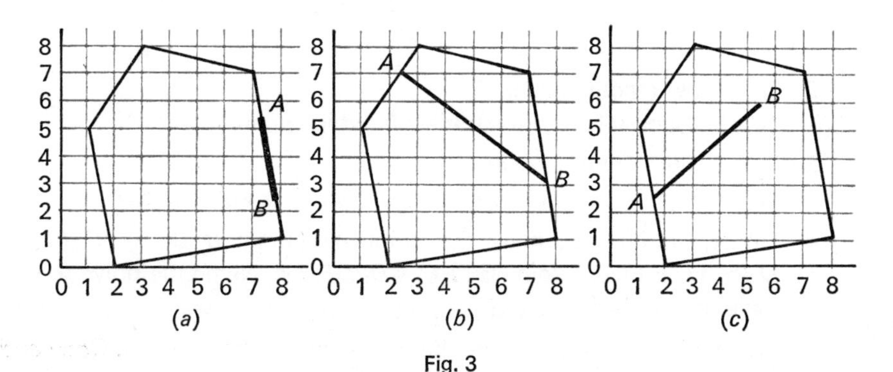

Fig. 3

(*c*) Copy and name the polygon shown in Figure 4. Can you find two points *A* and *B* of the polygon so that *AB* lies partly outside the polygon?

If, for every pair of points of the polygon, the line segment joining them lies entirely within the polygon, then the polygon is said to be *convex*.

The pentagon in Figure 2 is convex; the quadrilateral in Figure 4 is not convex.

(*d*) Look at Figure 1. Which of these polygons are convex?

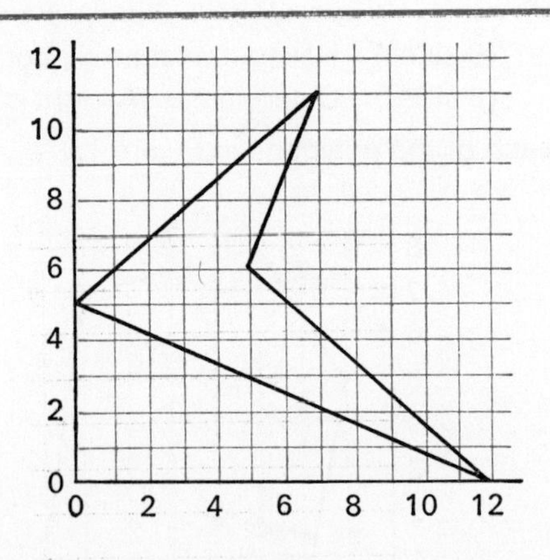

Fig. 4

Exercise A

1 Give five examples of polygonal frames and draw diagrams to illustrate them. For example, the bicycle frame shown in Figure 5.

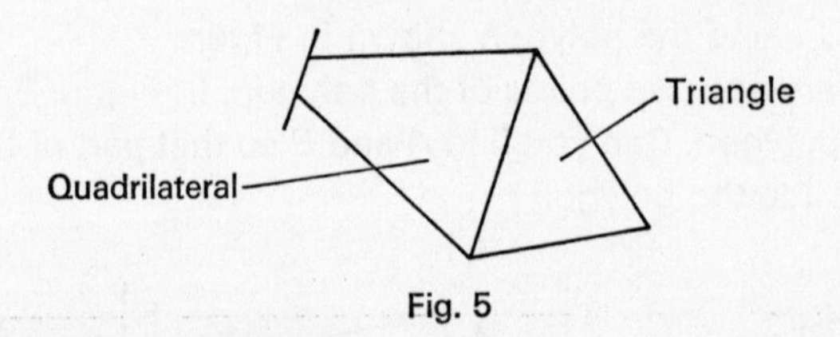

Fig. 5

2 (*a*) Draw a convex triangle.
(*b*) Is it possible to draw a triangle which is not convex?

3 (*a*) Draw a convex hexagon.
(*b*) Is it possible to draw a hexagon which is not convex?

4 On squared paper, mark the points *A* (1, 1), *B* (2, 5), *C* (4, 4), *D* (5, 1), *E* (3, 2). Join *AB*, *BC*, *CD*, *DE*, *EA*. Name the polygon you have drawn. Is it convex?

5 The angles of a polygon are the angles *inside* the polygon. Copy and complete the following table for the polygons in Figure 1.

Figure	Number of reflex angles	Convex yes or no
a	0	Yes
b	1	No
c		
d		
e		
f		
g		

Can you draw a convex polygon which has a reflex angle?

Exercise A

Questions 6, 7 are intended to make pupils aware that there are some limitations on the possible sizes of the interior angles of a particular polygon. Pupils are expected to obtain their answers by trial and error.

1 Steel bridges, cranes, tent frames, climbing frames, swings.

2 All triangles are convex.

3 (*a*) See Figure 1 (*c*); (*b*) see Figure 1 (*f*).

4 An opportunity to revise coordinates. *ABCDE* is a pentagon; it is not convex.

5

Figure	Number of reflex angles	Convex yes or no
a	0	Yes
b	1	No
c	0	Yes
d	0	Yes
e	4	No
f	1	No
g	1	No

No; each angle of a convex polygon is smaller than 180°.

6 (*a*) See Figure 1 (*a*); (*b*), (*c*) see Figure B; (*d*) impossible.

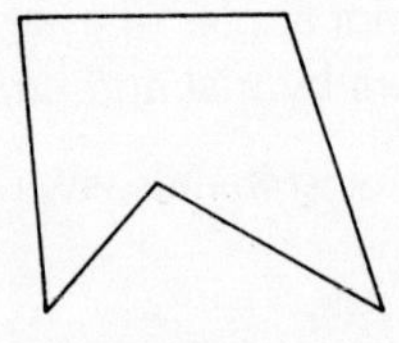

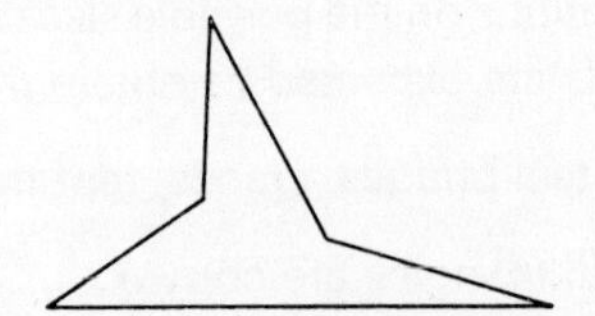

Fig. B

7 See Figure C (i). It is impossible to draw a convex or re-entrant hexagon with six right-angles. It is not necessary to consider crossed-over polygons (see Figure C (ii)) unless their existence is suggested by a pupil.

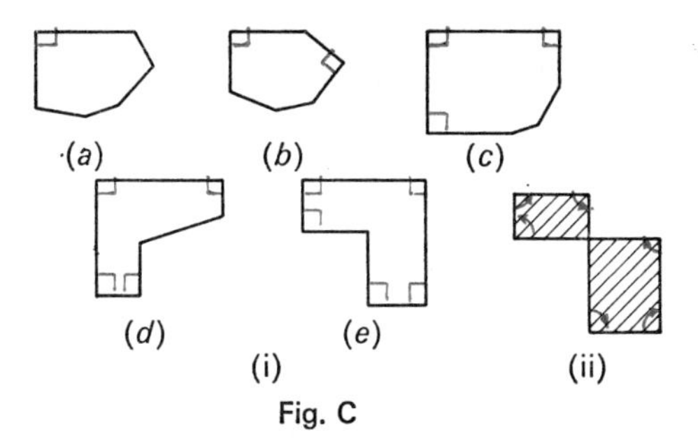

Fig. C

The concept of a polygon as a region cannot be applied to crossed-over polygons: they cannot be said to have interior or exterior angles. The sum of their angles has to be obtained by considering the angles turned through when moving around the boundary of the polygon making the rotation always in one direction. In this way, the sum of the angles of the crossed-over hexagon of Figure C (iii) can be seen to be zero.

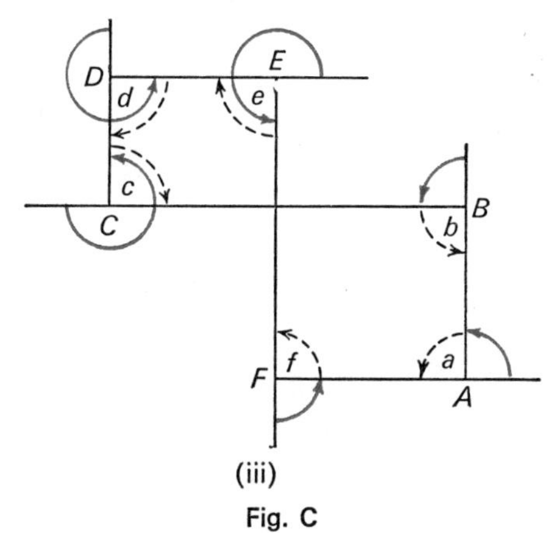

(iii)
Fig. C

The sum of the marked turns at the vertices is 6 half-turns. The sum of the angles through which rotations are made as you move round the polygon (marked with whole lines) is 3 turns. The sum of the angles between the sides of the polygon (marked with broken lines: the rotation from new to old sides) is $(a+\ldots+f)$. So

$$3 \text{ turns} + (a+\ldots+f) = 6 \text{ half-turns}.$$

$$a+\ldots+f = 0.$$

2. ANGLES OF POLYGONS

The main purpose of this section is to become familiar with some of the angle properties of polygons.

(*a*) *AB* and *AE* are sides of the polygon. A diagonal is a straight line joining any two non-adjacent vertices.

(*b*)

Convex polygon	Number of diagonals from one vertex	Number of cells	Shape of each cell
Quadrilateral	1	2	Triangular
Pentagon	2	3	Triangular
Hexagon	3	4	Triangular
Octagon	5	6	Triangular

(*c*) Yes; 180°; 540°.

(*d*) The sum of the interior angles of a convex hexagon is 720°.

(*e*) The sum of the interior angles of a convex octagon is 1080°.

6 Draw, if possible, a pentagon which has:

(*a*) no reflex angles; (*b*) one reflex angle;
(*c*) two reflex angles; (*d*) three reflex angles.

7 How many angles has a hexagon?
Sketch, if possible, hexagons which have

(*a*) one right-angle; (*b*) two right-angles;
(*c*) three right-angles; (*d*) four right-angles;
(*e*) five right-angles; (*f*) six right-angles.

2. ANGLES OF POLYGONS

(*a*) Sketch a convex pentagon like the one in Figure 6. A point where two sides meet is called a *vertex*. Draw lines from *A* to all the other vertices except *B* and *E*. These are the *diagonals* from *A*. Why are *AB* and *AE* not called diagonals?

How many diagonals from *A* can you draw? Into how many cells is the pentagon divided? What shape are these cells?

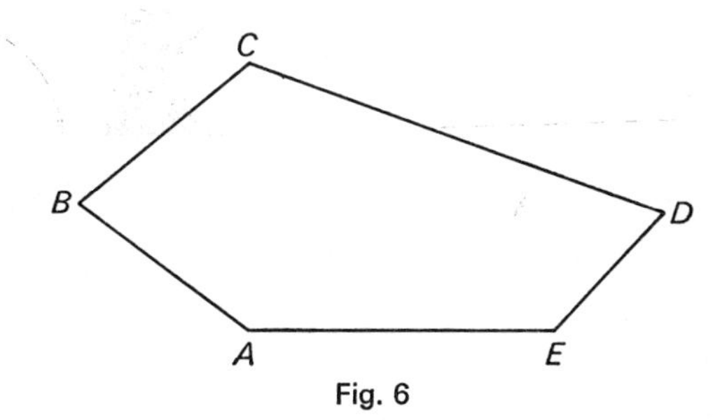

Fig. 6

(*b*) Copy and complete the following table.

Convex polygon	Number of diagonals from one vertex	Number of cells	Shape of each cell
Quadrilateral			
Pentagon	2	3	Triangular
Hexagon			
Octagon			

(*c*) The pentagon is divided into 3 triangular cells (see Figure 7). Do the angles of the 3 triangles make up the angles of the pentagon? What is the sum of the angles of a triangle? What is the sum of the angles of the pentagon?

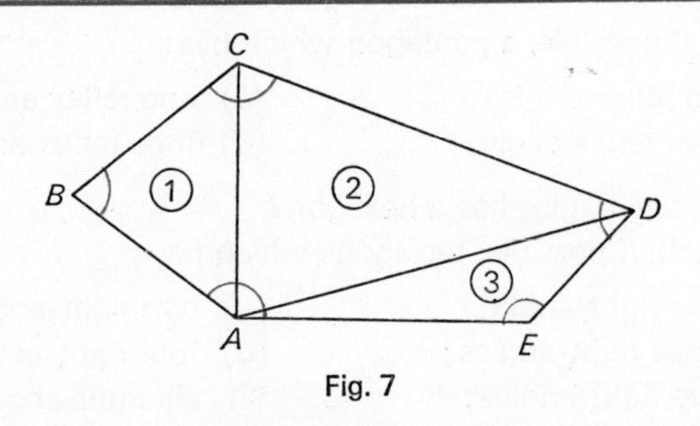

Fig. 7

(*d*) Draw a convex hexagon. Draw all the diagonals through one of the vertices. How many triangular cells are there? What is the sum of the angles of the hexagon?

(*e*) Use the method of (*c*) and (*d*) to find the sum of the angles of a convex octagon.

Experiment 1

Equipment: chalk or rope, a blackboard protractor.

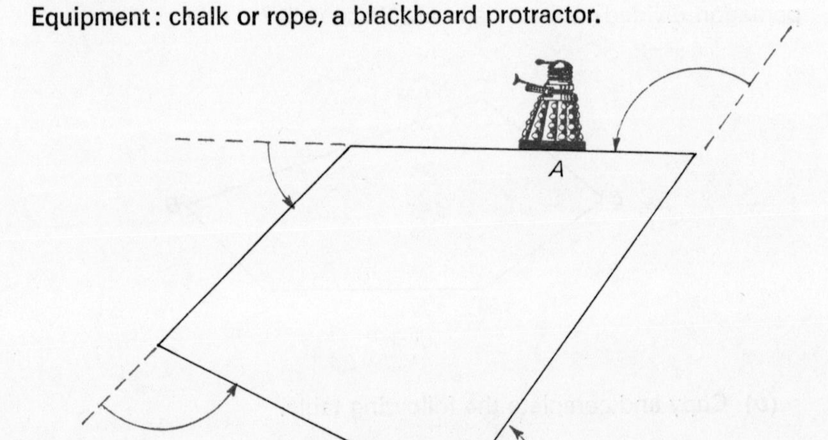

Fig. 8

Using chalk (or rope) mark out a large convex quadrilateral on the playground (or classroom floor). Start at *A* and walk once round the boundary of the quadrilateral turning always to the left. Ask a friend to measure the angle through which you turn at each vertex. What is the sum of these angles? Record your result.

Repeat this experiment for other convex polygons.

What happens when the polygon is not convex?

Experiment 1

This is suitable for groups of about four pupils. There may be practical difficulties and pupils should be encouraged to devise their own methods of overcoming these.

The sum of the angles turned through in walking once round the boundary of any convex polygon is 360°. This result remains true for re-entrant polygons (see Figure D). It is essential, however, that pupils remember to take account of the direction of the angles through which they turn and that they never turn through a reflex angle, as this would mean turning back on themselves.

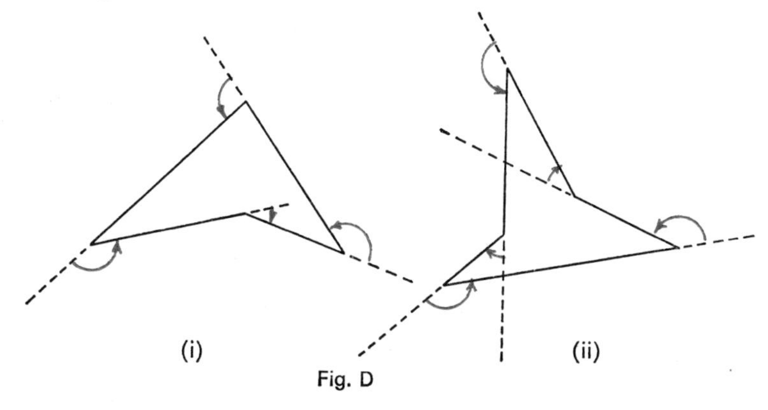

Fig. D

The sum of the angles turned through in walking round the boundary of a crossed-over polygon is a multiple of 360°. For example, in the case of the regular pentagram (see Figure E) the sum is $5 \times 144° = 720°$.

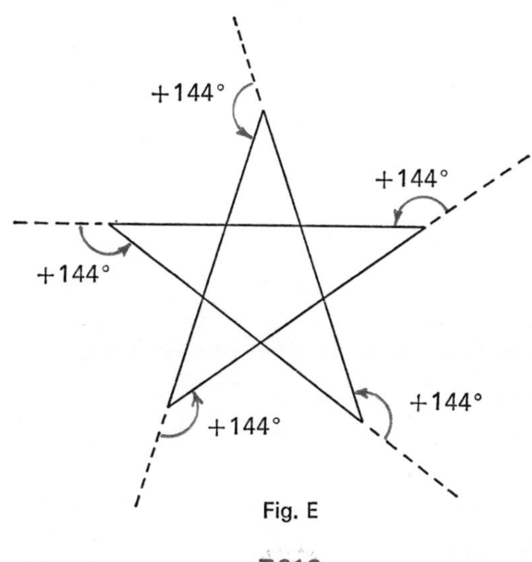

Fig. E

Experiment 2

The sum of the angles of any convex or re-entrant quadrilateral is 360°.

The authors have found that even less able pupils can appreciate Example 1 provided that Experiments 1 and 2 have previously been carried out with care. It is helpful if the first method is explained by the teacher with the aid of coloured chalks.

Experiment 2

Equipment: scissors, gummed paper.

Draw a convex quadrilateral on the gummed paper and cut it out. Mark the angles of the quadrilateral and tear off each corner. Place the four corners of the quadrilateral together (see Figure 10) and stick them into your exercise book.

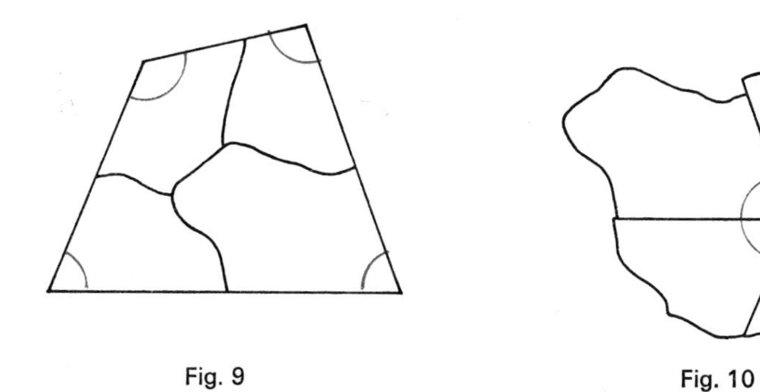

Fig. 9

Fig. 10

What can you say about the angles of your quadrilateral? Does your result depend upon the shape of the quadrilateral which you draw?

Example 1

Find the sum of the angles of the convex quadrilateral in Figure 11. All angle measurements are in degrees.

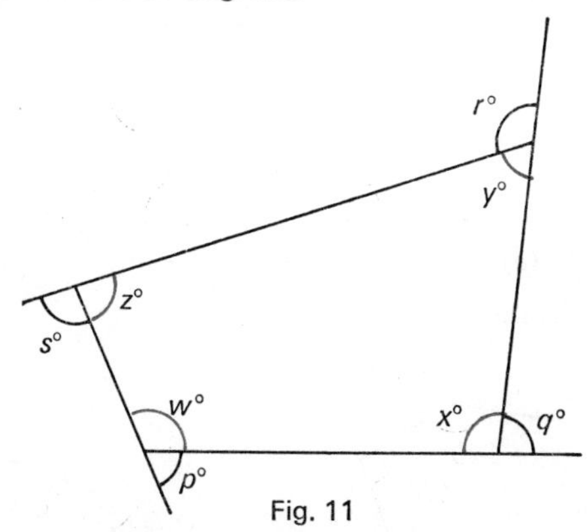

Fig. 11

First method. Imagine that you walk along the boundary of the quadrilateral. When you return to the starting point you have rotated through one complete turn.

At each corner you turn through the black angle; the four black angles together form one whole turn or 360°.

Each black angle together with its red angle adds up to 180°.

$$4 \text{ black angles} + 4 \text{ red angles} = 4 \times 180° = 720°$$

$$4 \text{ red angles} = 720° - 4 \text{ black angles}$$

$$= 720° - 360°$$

$$= 360°.$$

The sum of the angles inside a quadrilateral is 360° or one complete turn.

Second method. Draw a diagonal from one of the vertices. There are now two triangular cells and the sum of the angles of each is 180°. The sum of the angles of the quadrilateral is twice this amount, that is 360° (see Section 2 (*c*)).

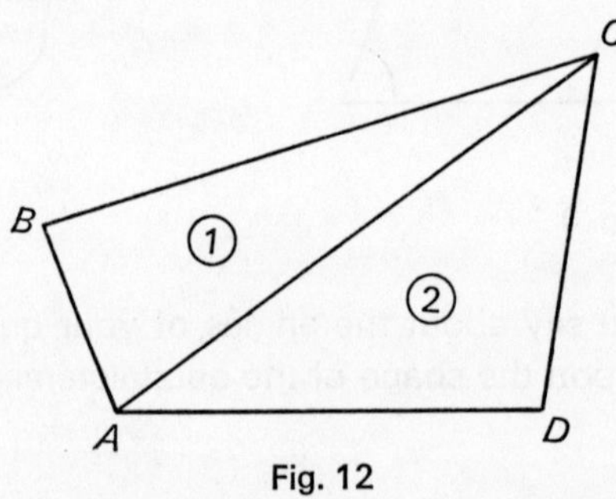

Fig. 12

Exercise B

1 Find the angles represented by small letters in Figure 13.

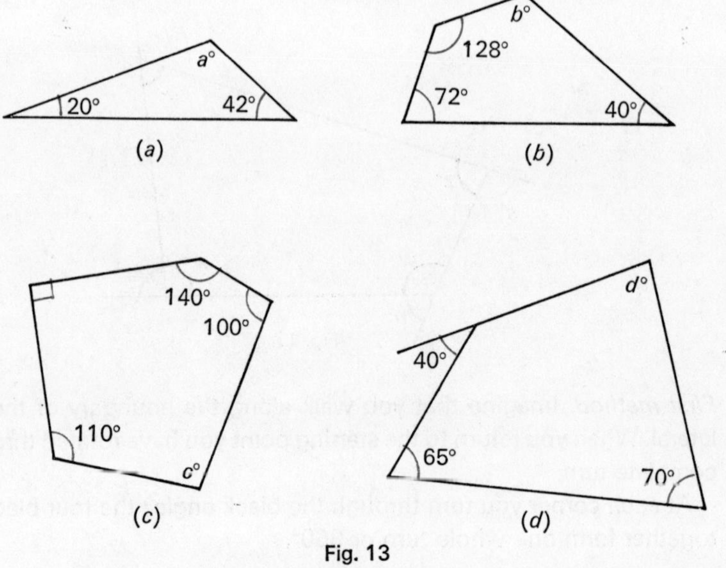

Fig. 13

Exercise B

In giving the answers to this exercise, crossed-over polygons have not been considered.

1 (*a*) 118°; (*b*) 120°; (*c*) 100°; (*d*) 85°.

2 Yes. A re-entrant quadrilateral can be divided into two triangular cells (see Figure F). If the question is answered with the aid of a protractor, then there may be some difficulty over which angle at P should be measured.

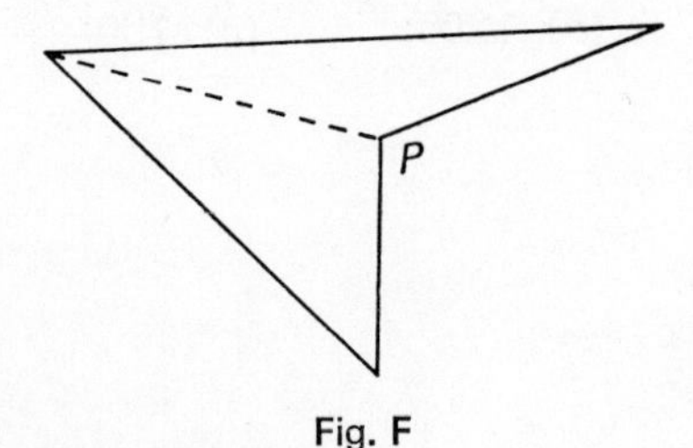

Fig. F

3 60°.

4 (*a*) Hexagon; 8 right-angles or 720°.
(*b*) Octagon; 12 right-angles or 1080°.
(*c*) Pentagon; 6 right-angles or 540°.
(*d*) Quadrilateral; 360°.
(*e*) Octagon; 12 right-angles or 1080°.
(*f*) Octagon; 12 right-angles or 1080°.

The polygons shown in Figure 14 (*a*), (*b*), (*e*) are re-entrant but this does not affect the sum of their interior angles.

5 540°.

6 100°.

7 900°.

8 1800°.

9 (*a*) No;
(*b*) yes; the angles of a polygon add to a multiple of 180°.

10 Yes.

11 8 sides; 8 vertices. $1080° = 6 \times 180°$ and an octagon can be divided into six triangluar cells.

2 Draw a quadrilateral with a reflex angle. This is called a 're-entrant' quadrilateral. Is the sum of the angles still 360°?

3 Three angles of a quadrilateral are each 100°. What size is the fourth angle?

4 Name the polygons in Figure 14 and find the sum of their angles. (Be careful to use the angles which are *inside* the polygons. Do not use a protractor.)

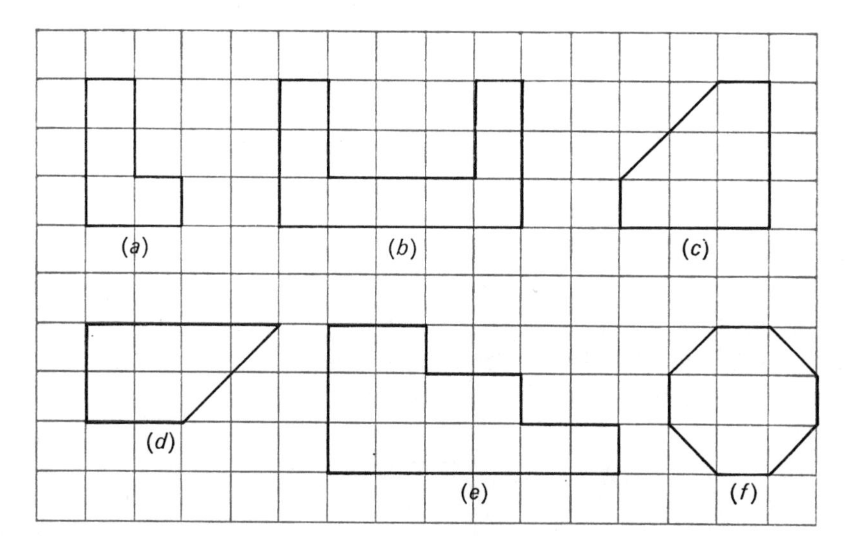

Fig. 14

5 Use the first method of Example 1 to find the sum of the angles of a convex pentagon.

6 Four angles of a pentagon are each 110°. What size is the fifth angle?

7 Find the sum of the angles of a seven-sided polygon (a heptagon).

8 Find the sum of the angles of a twelve-sided figure (a dodecagon).

9 Is it possible to draw a polygon with angles which add up to (*a*) 450°; (*b*) 900°?

10 The angles of a polygon add up to 360°. Must the polygon be a quadrilateral?

11 The angles of a polygon add up to 1080°; how many sides has it? How many vertices has it?

SUMMARY

A polygon is a plane figure with straight sides.
Here are the names of some important ones:

3 sides	triangle;
4 sides	quadrilateral;
5 sides	pentagon;
6 sides	hexagon;
8 sides	octagon.

A polygon is convex if, for every pair of points within it, the line segment joining those points lies entirely within the polygon.

The sum of the angles of a quadrilateral is 360°. (Remember that the 'angles of a quadrilateral' means 'the angles inside the quadrilateral'.)

3. REGULAR POLYGONS

When a polygon has all its sides equal *and* all its angles equal, it is called *regular*. For example, the regular quadrilateral is the square.

(*a*) The regular triangle also has a special name. What is it?

(*b*) In Figure 15, which polygons are regular?

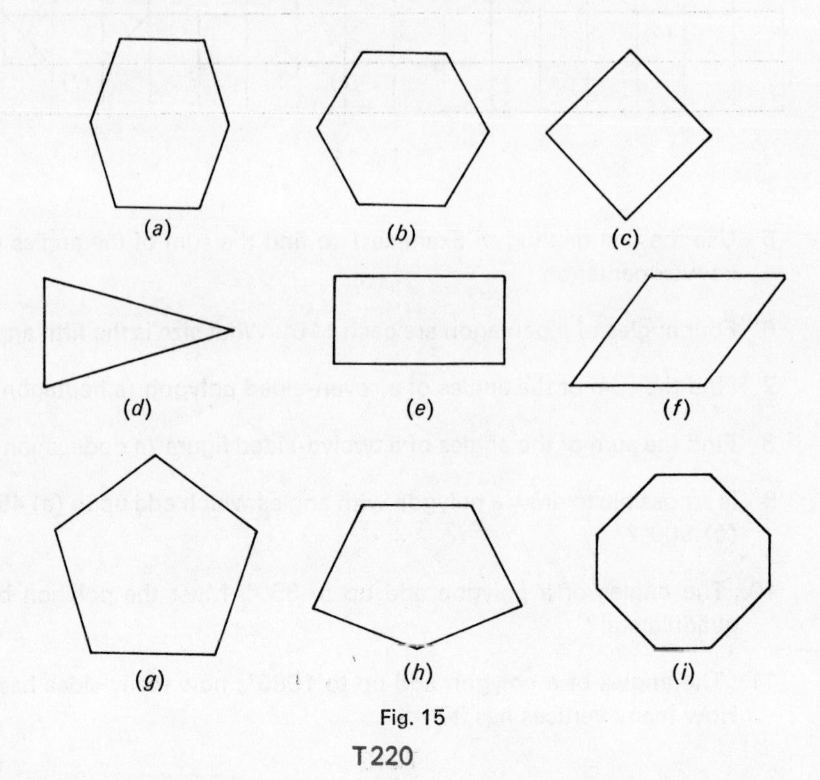

Fig. 15

3. REGULAR POLYGONS

(*a*) Equilateral triangle.

(*b*) (*b*), (*c*), (*g*). It is important that pupils should realize that a regular polygon has all its angles equal as well as all its sides equal.

Figure 15 shows:

(*a*) a hexagon with equal sides;
(*b*) a regular hexagon;
(*c*) a regular quadrilateral or square;
(*d*) an isosceles triangle;
(*e*) a quadrilateral with equal angles (rectangle);
(*f*) a quadrilateral with equal sides (rhombus);
(*g*) a regular pentagon;
(*h*) a pentagon with equal sides;
(*i*) an octagon with equal angles.

(c)

Figure	a	b	c	d	e	f	g	h	i
Number of lines of symmetry	2	6	4	1	2	2	5	1	4
Order of rotational symmetry	2	6	4	1	2	2	5	1	4

(d) 360°; 90°.

(e) 60°.

(f) The angles are all equal and, since they add to 360°, each is 72°.

Exercise C

1 $\angle BOC = 72°$; isosceles; $\angle OBC = 54° = \angle OCB = \angle OBA$; 108°.

2 $\angle COD = 72° = \angle DOE = \angle EOA = \angle AOB = \angle BOC$. The pentagon can be drawn by making five angles of 72° at the centre of the circle.

3 $\angle AOF = 60°$; equilateral; $\angle OAF = 60° = \angle OFA = \angle OFE$; 120°.

4 Equilateral triangles. Some pupils may use the method suggested in Question 2. The last part of the question is designed to provoke discussion leading to the method of construction shown in Figure G. The compasses are kept open at the radius of the circle and successive arcs are drawn round the circle. The last arc should pass through the starting point.

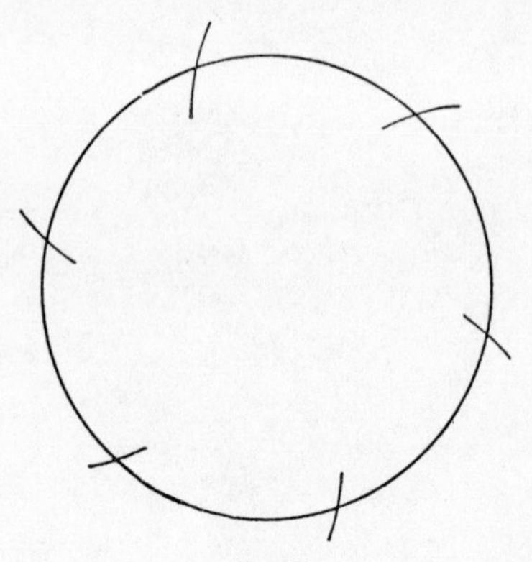

Fig. G

(*c*) Comment on the symmetries of these polygons.

(*d*) What is the sum of the angles of a regular quadrilateral? Work out the number of degrees in each angle.

(*e*) How many degrees are there in each angle of a regular triangle?

(*f*) When a polygon is regular it fits exactly into a circle. Figure 16 shows a regular pentagon fitting exactly into a circle with centre *O*. How many degrees are there in $\angle BOC$, $\angle COD$, $\angle DOE$, $\angle EOA$, and $\angle AOB$?

Exercise C

1 In Figure 16, *O* is the centre of the circle. Find $\angle BOC$.
What kind of triangle is *BOC*?
What are the sizes of $\angle OBC$, $\angle OCB$ and $\angle OBA$?
How many degrees are there in an angle of a regular pentagon?

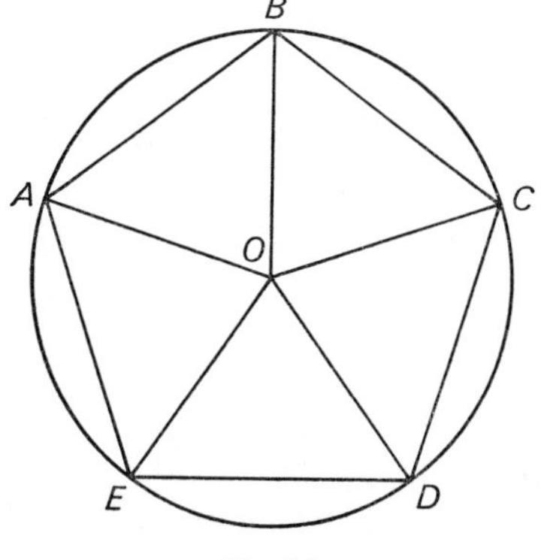

Fig. 16

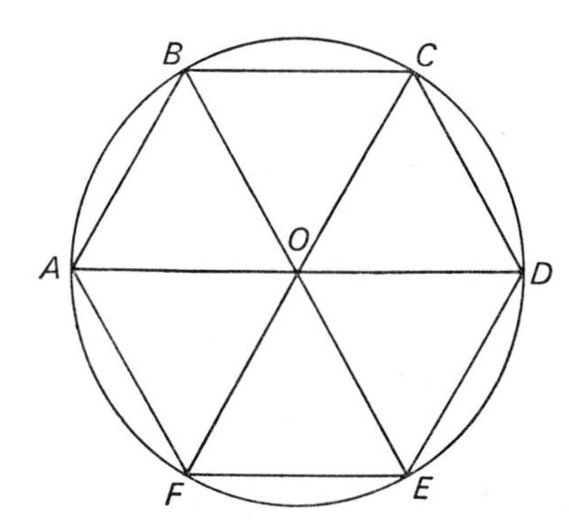

Fig. 17

2 Look at Figure 16. What is the size of each of the angles at the centre *O*?
By starting with a circle of radius 6 cm, make an accurate drawing of a regular pentagon.

3 In Figure 17, *ABCDEF* is a regular hexagon in a circle whose centre is *O*. Find the size of $\angle AOF$.
What kind of triangle is *AOF*?
Find the sizes of $\angle OAF$, $\angle OFA$ and $\angle OFE$.
How many degrees are there in an angle of a regular hexagon?

4 Look at Figure 17. The regular hexagon is divided into six triangles. What kind of triangles are they? By starting with a circle of radius 5 cm, make an accurate drawing of a regular hexagon.
Can you draw a regular hexagon without using a protractor?

5 Draw a large regular pentagon by first drawing a circle. Draw all the diagonals and you will find a smaller regular pentagon inside the first one. Draw the diagonals of the smaller pentagon and repeat the process as many times as you can.

6 How many degrees are there in an angle of a regular octagon?

7 Copy and complete the following table about convex polygons. The 'number of triangles' is the number which are formed when all the diagonals are drawn from *one* vertex.

Number of sides	Number of triangles	Sum of all angles	Angle of regular polygon
3	1	180°	$\frac{180°}{3} = 60°$
4	2	$2 \times 180° = 360°$	$\frac{360°}{4} = 90°$
5			
6	4	$4 \times 180° = 720°$	$\frac{720°}{6} = 120°$
7			
8			
9			
10			
12			
n			

8 (*a*) Is an angle of a regular octagon larger or smaller than the angle of a regular pentagon?
(*b*) Is an angle of a regular 40-sided polygon larger or smaller than the angle of a regular 20-sided polygon?

D 9 (*a*) What is the smallest possible angle of a regular polygon? Which regular polygon has angles of this size?
(*b*) What is the smallest angle which is too big to be an angle of a regular polygon?

10 Is it possible for the angle of a regular polygon to be: (*a*) 90°; (*b*) 75°; (*c*) 108°? Give reasons for your answers.

11 The angle of a regular polygon is 162°. How many sides has it?
(*Hint*: find what angle you would turn through at each vertex in walking round it.)

12 Take a strip of paper about 25 cm long and 2·5 cm wide. Tie a knot in it (see Figure 18*a*) and pull it tight (see Figure 18*b*).

5 See Figure H.

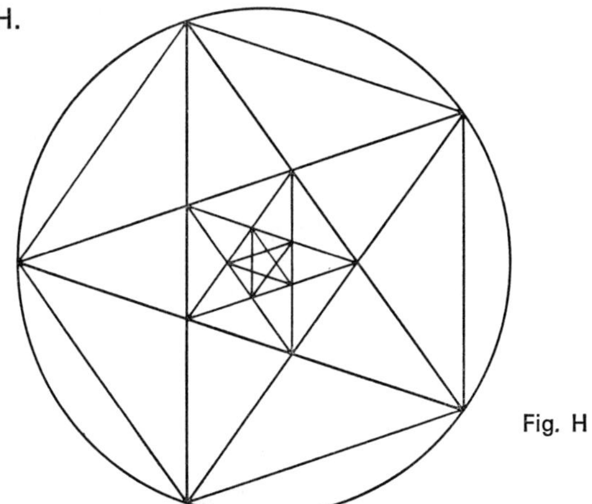

Fig. H

6 Formulae for finding angles are unnecessary. It should be possible to work from first principles in each case. The method suggested in Question 1 can be used. 135°.

7 Some pupils may need to look again at the table on Page 103 before attempting this question.

Number of sides	Number of triangles	Sum of all angles	Angle of regular polygon
3	1	180°	$\frac{180°}{3}=60°$
4	2	$2\times180°=360°$	$\frac{360°}{4}=90°$
5	3	$3\times180°=540°$	$\frac{540°}{5}=108°$
6	4	$4\times180°=720°$	$\frac{720°}{6}=120°$
7	5	$5\times180°=900°$	$\frac{900°}{7}=128\frac{4}{7}°$
8	6	$6\times180°=1080°$	$\frac{1080°}{8}=135°$
9	7	$7\times180°=1260°$	$\frac{1260°}{9}=140°$
10	8	$8\times180°=1440°$	$\frac{1440°}{10}=144°$
12	10	$10\times180°=1800°$	$\frac{1800°}{12}=150°$
n	$n-2$	$(n-2)\times180°=180(n-2)°$	$\frac{180(n-2)°}{n}$

The last entry of the table may prove hard for some pupils. If this is so, it may be delayed until later in the course.

8 The table in Question 7 shows that the size of the interior angle of a regular polygon increases with the number of sides.

(*a*) Larger; (*b*) larger.

9 The table in Question 7 helps here.

(*a*) 60°, equilateral triangle.

(*b*) As the number of sides of a regular polygon increases, the size of the angle of the polygon also increases and approaches the value 180°. Thus there is no largest possible angle but the size cannot increase beyond 180°. There is an 'upper bound' of 180°.

10 This question is intended to strengthen the concept that the size of the angle of a regular polygon increases with the number of sides. Answers such as:

'(*a*) Yes, square;

(*b*) no, a polygon cannot have three and a bit sides;

(*c*) yes, pentagon'

are acceptable.

11 The size of the angle turned through at each vertex is 18°.
$360 \div 18 = 20$. Therefore the polygon has 20 sides.

12 Regular pentagon. Machine rolls (of various widths) can be obtained from office stationers and are useful for knotting polygons.

Most pupils are able to obtain a square by paper-folding. A method of paper-folding an equilateral triangle appears in the text on page 130. See also *Experiments in Mathematics*, Stage 1, p. 17, Pearcy and Lewis.

13 Seven. Regular hexagon.

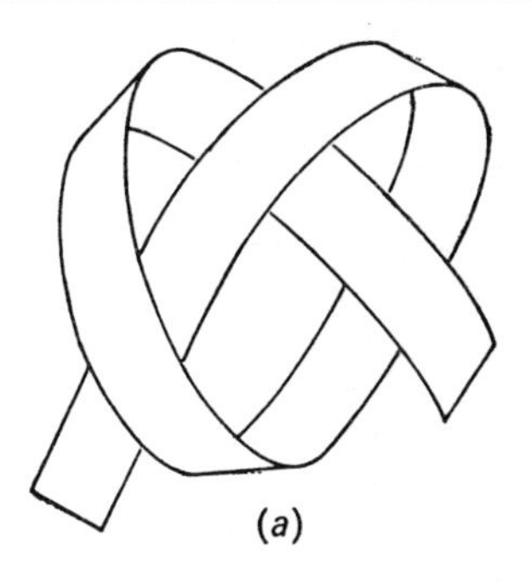

(a)

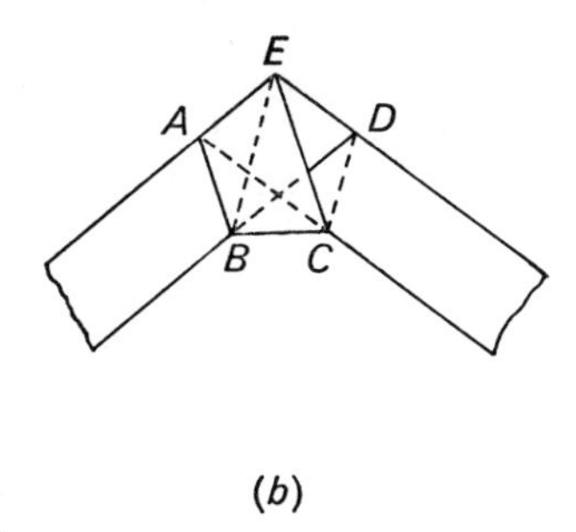

(b)

Fig. 18

Flatten your knot very carefully.
What shape is *ABCDE*?
Can you obtain any other regular polygons by folding paper?
How?

13 Fit pennies together flat on a table so that one is surrounded by a ring made of the others. How many pennies are there altogether? What shape is obtained by joining with straight lines the centres of the pennies of the outer ring?

SUMMARY

A regular polygon has all its sides equal *and* all its angles equal.

A regular polygon fits exactly into a circle (that is, a circle can be drawn to pass through all the vertices of the polygon).

A regular triangle is an equilateral triangle.

A regular quadrilateral is a square.

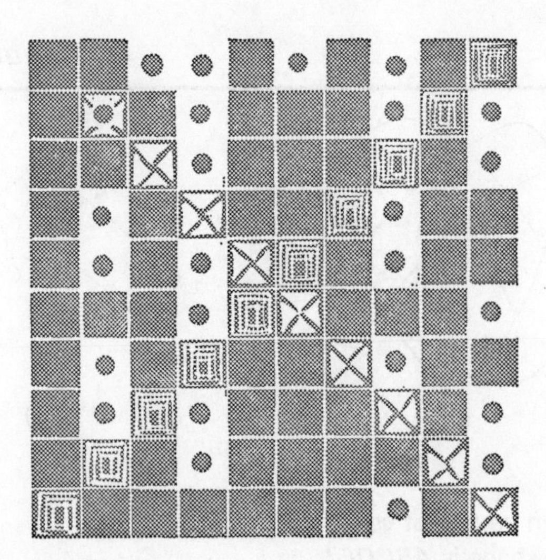

From 13

8. Further number patterns

1. MULTIPLES AND COMMON MULTIPLES

The following numbers have all been expressed in terms of two of their factors:

$$6 = 2\times3 \qquad 12 = 4\times3 \qquad 18 = 6\times3$$

$$9 = 3\times3 \qquad 15 = 5\times3 \qquad 21 = 7\times3$$

Each of these numbers has 3 as one of its factors and so we say that these rectangle numbers are all *multiples* of 3. They are a few of the members of the set of multiples of 3.

{multiples of 3} = {3, 6, 9, 12, 15, 18, 21, 24, . . .}.

This set is not like the others you have met because there are many more multiples of three than we have shown. In fact we can go on multiplying numbers by three for as long as we like, and you will always get new numbers that are members of the set.

In the same way:

{multiples of 7} = {7, 14, 21, 28, 35, 42, . . .}.

1.1 Multiple patterns

Rule out a square with eleven spaces across and eleven spaces down. Number the top and left-hand edges as in the diagram. This shows the units and the tens digits for the numbers.

8. Further number patterns

The contents of this chapter follow immediately from that of Chapter 1. We have discussed sets of factors and now progress to sets of multiples. Though this does introduce infinite sets, it is not necessary to name or define them at this stage. We shall use the notation of three dots '...' to indicate that the sequence of numbers continues.

This chapter also extends the notation we have used in association with sets. The relation of 'inclusion' is introduced. It should be realized that this can only apply to individual members (or elements) of the set. It is not the same as the relation 'is a subset of' for example:

$$10 \in \{\text{triangle numbers}\},$$

$$\{10\} \notin \{\text{triangle numbers}\}.$$

The sign for the operation of intersection is also introduced. Notice that when two sets intersect, their intersection is another set; it is the set of elements that are members of both the given sets.

1. MULTIPLES AND COMMON MULTIPLES

1.1 Multiple patterns

These multiple patterns should be constructed by the pupils themselves, for it is in the process of construction that they realize how the patterns develop. These patterns should be retained for later reference.

Further number patterns

Pupils must take great care in marking the multiples of 2, 3, 4, etc., as one mistake will obviously spoil the whole pattern.

14 $\in$ {multiples of 4} is the only statement which is not true.

{45, 90, 135, . . .} = {multiples of 9} $\cap$ {multiples of 5}.

Units digits

Tens digits

	0	1	2	3	4	5	6	7	8	9
0										
1										
2										
3										
4										
5										
6										
7										
8										
9										

Put a 2 (or some 2-mark, for example, a coloured dot) in the space for 2 and in every second space. This marks the multiples of 2.

Put a 3 (or some 3-mark) in the space for 3 and in every third space.

Put a 4 (or a 4-mark) in the space for 4 and in every fourth space. As $4 = 2 \times 2$, make the 4-mark a repeated 2-mark. As $6 = 2 \times 3$, you will find a 2-mark and a 3-mark in the space for 6. So let this pair form the 6-mark.

Continue at least as far as multiples of 11. Invent new marks for 5, 7 and 11. How are you going to mark 8, 9 and 10? Notice the patterns. Which way do the lines of 3-marks slope? Are the lines of 5-marks across or down the page? Where do the lines of 9-marks cross the lines of 7-marks?

We sometimes use the symbol '$\in$' to stand for 'is a member of'.

$$27 \in \{\text{multiples of } 3\}$$

is read '27 is a member of the set of multiples of 3'.

Are the following true:

$$14 \in \{\text{multiples of } 4\},$$

$$64 \in \{\text{square numbers}\},$$

$$75 \in \{\text{rectangle numbers}\}?$$

Sometimes numbers are members of several different sets. On your square of multiples, do the sloping lines of 9-marks cross the column of 5-marks? Is it true to say that

$$45 \in \{\text{multiples of } 9\},$$

and that $$45 \in \{\text{multiples of } 5\}?$$

What other numbers can be used instead of 45? These two sets have some members in common; they are said to *intersect*. (See page 25.)

Further number patterns

To remind you of the special sign for intersection:

Let N stand for the set of multiples of 9 and F stand for the multiples of 5. We can write

$$N = \{\text{multiples of } 9\}, \quad F = \{\text{multiples of } 5\}.$$

The intersection of these sets is written $N \cap F$ and this is read 'N intersection F'.

Exercise A

Use the number square to help you to answer these questions. In all the questions, A, B and C refer to the sets defined in Question 1.

1 List the members of the following sets:

$$A = \{\text{multiples of 2 which are less than 40}\},$$

$$B = \{\text{multiples of 3 which are less than 40}\},$$

$$C = \{\text{multiples of 5 which are less than 40}\}.$$

2 Copy the following statements putting in 'is' or 'is not' where you see the space marked ——:

(*a*) 2 —— a member of A;
(*b*) 20 —— a member of B;
(*c*) 20 —— a member of C.

3 Using the symbol $\in$, write down the set or sets (A, B or C) to which the following numbers belong. (Notice that $15 \in B$ and $15 \in C$.)

(*a*) 32; (*b*) 27; (*c*) 28; (*d*) 35; (*e*) 10; (*f*) 18; (*g*) 20; (*h*) 45; (*i*) 40; (*j*) 36; (*k*) 30; (*l*) 42.

4 Is it true that $20 \in A \cap C$?

5 Find two other members of $A \cap C$.

6 Make a list of all the members of $A \cap B$.

7 List the members of $B \cap C$.

8 Which number is a member of A, B and C, that is of $A \cap B \cap C$?

9 The set $A \cap B$ is the set of numbers less than 40 which are multiples of 2 and 3. This is {common multiples of 2 and 3 which are less than 40}.
Copy and complete the following statement:

$B \cap C$ = {common multiples of . . . and . . . which are less than. . .}.

Exercise A

1 $A = \{2, 4, 6, 8, 10, 12, 14, 16, 18, 20, 22, 24, 26, 28, 30, 32, 34, 36, 38\}$;
$B = \{3, 6, 9, 12, 15, 18, 21, 24, 27, 30, 33, 36, 39\}$;
$C = \{5, 10, 15, 20, 25, 30, 35\}$.

2 (*a*) Is; (*b*) is not; (*c*) is.

3 (*a*) $32 \in A$; (*b*) $27 \in B$; (*c*) $28 \in A$; (*d*) $35 \in C$;
(*e*) $10 \in A$, $10 \in C$; (*f*) $18 \in A$, $18 \in B$;
(*g*) $20 \in A$, $20 \in C$;
(*h*) 45 is not a member of any of the given sets;
(*i*) 40 is not a member of any of the given sets;
(*j*) $36 \in A$, $36 \in B$;
(*k*) $30 \in A$, $30 \in B$, $30 \in C$;
(*l*) 42 is not a member of any of the given sets.

4 Yes.

5 10, 30.

6 $A \cap B = \{6, 12, 18, 24, 30, 36\}$.

7 $B \cap C = \{15, 30\}$.

8 $30 \in A \cap B \cap C$.

9 $B \cap C = \{$common multiples of 3 and 5 which are less than 40$\}$.

2. PRIME NUMBERS

It is fairly important to remind people of the first few prime numbers but there is no need to learn them by heart. A pupil should be prepared to test numbers less than 100 for factors to discover the primes where necessary.

The irregularity of prime numbers interests some pupils enough to make them spend a great deal of time searching either for a formula for primes or at least for some regular property. For example, once it is realized that, in general, prime numbers become less frequent as one moves along the sequence of numbers, it might be thought that there is a largest prime number. That this is not the case can easily be proved by the following argument:

Let P be the largest prime number. Consider the number

$$(2\times3\times5\times7\times11\times13\times\ldots\times P)+1.$$

This number is greater than P and is either a prime or not. If it is a prime, then there is a prime greater than P. If it is not a prime then it must have a prime factor. This prime is not any of the primes up to P because, for example, 13 is a factor of the product but there is a remainder of 1 when this number is divided by 13. So the prime factor must be a prime greater than P. So, in either case, whether the new number is prime or not, we have shown that there must exist a prime greater than P. This contradicts our assumption that P is the largest prime.

Formulae giving prime numbers might also be worth investigating. For example, x^2+x+41 (for $x<40$) and x^2+x+17 (for $x<16$) and $2^{2^n}+1$ (for $n<5$). Although it can be proved that no formulae such as these exist that would give all prime numbers, it is still possible that a formula could be found to give some of the prime numbers without ever giving a number which is not prime. A fuller treatment of primes is given in Mansfield and Thompson, *Mathematics, a New Approach*, Book 1, Chapter 5, or, for a more advanced treatment, *Solved and Unsolved Problems in Number Theory*, by Daniel F. Shank, or *The Theory of Numbers*, by Hardy and Wright.

2.1 Prime factors

It is best to be systematic when searching for prime factors. Start with the smallest prime. Any prime factor of a number must be smaller than the square root of the number and this will give a rough guide as to when one may give up the search.

No element may be repeated when listing the elements of a set. It is for this reason we say that the set of prime factors of, for example, 12 is {2, 3}.

2. PRIME NUMBERS

Prime numbers are very special and important numbers. In a way, they are the 'raw material' from which we make all other numbers. They are like the bricks which go into the building of a house or the paints with which we can make pictures.

Each prime number is bigger than 1 and has only two factors, 1 and itself. So, in a completed table of multiples, any space with just *one* mark must show a prime number.

(To find all the prime numbers up to 100, it might be best to draw a new number square and to use shading rather than number marks. Shade all the multiples of 2 except 2 itself. Look for the next empty square after 2. It is the 3 square. Shade all the multiples of 3 except 3 itself. Look for the next empty square after 3. It is the 5 square. Shade all the multiples of 5 except 5 itself a nd so on.)

Are there 25 prime numbers less than 100? Is there any regularity in the pattern of numbers between each prime number? Are there the same number of primes numbers between 1 and 50 as between 51 and 100? Would you expect there to be a biggest prime number?

2.1 Prime factors

Example 1

What are the prime factors of 30? Multiply them together.

$$30 = 2 \times 15 \quad \text{but} \quad 15 = 3 \times 5 \quad \text{and so} \quad 30 = 2 \times 3 \times 5.$$

Check with the multiple square and you will see that the square representing 30 has in it the marks for 2, 3 and 5.

$$\{\text{prime factors of } 30\} = \{2, 3, 5\}.$$

Answer the same question about the number 42.

Some of the spaces in our multiple pattern have a mark that is repeated. This shows that we sometimes need to use the same prime factor more than once.

Example 2

Find the prime factors of 12.

$$12 = 2 \times 6 \quad \text{but} \quad 6 = 2 \times 3 \quad \text{and so} \quad 12 = 2 \times 2 \times 3.$$

Although the only distinct prime factors of 12 are 2 and 3 we need to use the 2 twice before we can write down 12 in terms of its prime factors.

Notice that:

(*a*) {factors of 12} = {1, 2, 3, 4, 6, 12},
(*b*) {prime factors of 12} = {2, 3},
(*c*) $12 = 2 \times 2 \times 3$.

Example 3

Find the prime factors of 56 and write it as a product of primes.

This number is rather larger than the others we have looked at, so we must think of an orderly way to tackle it. Start by trying to divide it by the smallest prime number. If this divides into it a whole number of times, we know that it is a factor.

2)56
2)28 now try 2 again
2)14 and again
7 and we know that 7 is a prime number.

$$\{\text{prime factors of } 56\} = \{2, 7\},$$

$$56 = 2\times 2\times 2\times 7.$$

In the same way we can show that:

$$36 = 2\times 2\times 3\times 3,$$

$$75 = 3\times 5\times 5,$$

$$32 = 2\times 2\times 2\times 2\times 2.$$

This is a very long way to write out the answers, but we can shorten it.

When you first met square numbers you found that 7×7 could be written as 7^2. We can use the same idea again and write 32 as 2^5 which is read as '2 to the power of 5' or '2 to the fifth'.

In the same way $36 = 2^2\times 3^2$; $75 = 3\times 5^2$; and $56 = 2^3\times 7$.

Exercise B

1 Write each of the following as products of primes:

(*a*) 8; (*b*) 15; (*c*) 27;

(*d*) 54; (*e*) 63; (*f*) 72.

2.2 Finding prime numbers

Example 4

Is 221 a prime number?

To answer this question we have to find out whether 221 has any prime factors. This Is done by dividing by prime numbers to see whether there will be a remainder. Always start with the smallest prime and work your way through.

Exercise B

1 (*a*) $8=2^3$; (*b*) $15=3\times5$; (*c*) $27=3^3$;

(*d*) $54=2\times3^3$; (*e*) $63=3^2\times7$; (*f*) $72=2^3\times3^2$.

2.2 Finding prime numbers

There are, of course, divisibility tests which could be used, and the pupils will probably know the tests for two, three and five (ending in an even number; the sum of the digits is a multiple of three; ending in zero or five). Beyond these, the tests would seem to be of little value in this context because their complexity outweighs their usefulness.

Further number patterns

Exercise C

1 (*a*) $A = \{1, 3, 5, 15\}$,
$B = \{1, 2, 3, 6, 9, 18\}$,
$C = \{1, 2, 3, 4, 6, 12\}$;

(*b*) $A \cap P = \{3, 5\}$,
$B \cap P = \{2, 3\}$,
$C \cap P = \{2, 3\}$.

2 *A* has 8; *B* has 4; *C* has 5; *D* has 5; *E* has 3.

No. The number of primes between 100 and 120 is 5.

$F = \{101, 103, 107, 109, 113\}$.

3 (*a*) {prime factors of 20} $= \{2, 5\}$, $20 = 2^2 \times 5$;
(*b*) {prime factors of 27} $= \{3\}$, $27 = 3^3$;
(*c*) {prime factors of 48} $= \{2, 3\}$, $48 = 2^4 \times 3$;
(*d*) {prime factors of 231} $= \{3, 7, 11\}$, $231 = 3 \times 7 \times 11$;
(*e*) {prime factors of 455} $= \{5, 7, 13\}$, $455 = 5 \times 7 \times 13$.

4 (*c*) 139 and (*e*) 199.

5 $1000 = 2^3 \times 5^3$.

6 Larger. The difference is 24.

7 (*a*) 6600.
(*b*) This number will be different because any number is uniquely expressed as a product of primes and these products are composed of different factors.
(*c*) The first is larger.

$$\left.\begin{matrix}2\\3\\5\\7\\11\end{matrix}\right\}\text{is not a factor as there is a remainder after dividing 221 by}\left\{\begin{matrix}2\\3\\5\\7\\11\end{matrix}\right.$$

13 is a factor because it divides into 221 exactly 17 times.
So $221 = 13 \times 17$ and is therefore not a prime number.

Exercise C

1 (*a*) What are the members of the following sets?

$$A = \{\text{factors of } 15\},$$
$$B = \{\text{factors of } 18\},$$
$$C = \{\text{factors of } 12\}.$$

(*b*) If $P = \{\text{prime numbers}\}$, what are the members of:

$$A \cap P, \quad B \cap P, \quad C \cap P?$$

2 $A = \{\text{prime numbers between } 0 \text{ and } 20\}$,
$B = \{\text{prime numbers between } 20 \text{ and } 40\}$,
$C = \{\text{prime numbers between } 40 \text{ and } 60\}$,
$D = \{\text{prime numbers between } 60 \text{ and } 80\}$,
$E = \{\text{prime numbers between } 80 \text{ and } 100\}$.

How many members has each of the sets A, B, C, D and E? From these answers, can you say how many members there would be in the set F if $F = \{\text{prime numbers between } 100 \text{ and } 120\}$? Now find the actual number of primes between 100 and 120. List the members of this set.

3 Find the prime factors of the following numbers, then write them as products of primes:

(*a*) 20; (*b*) 27; (*c*) 48; (*d*) 231; (*e*) 455.

4 Which of the following numbers are prime?

(*a*) 324; (*b*) 927; (*c*) 139; (*d*) 161; (*e*) 199.

5 Write 1000 as a product of primes.

6 Is 2^{10} larger or smaller than 1000? Find the difference.

7 (*a*) Which number can be written:

$$2^3 \times 3 \times 5^2 \times 11?$$

(*b*) A different number can be written:

$$2^2 \times 3^2 \times 5 \times 11.$$

Without working out the second number can you say how you know that it will be different?

(*c*) Which of the two numbers is the larger?

8 A quick test to discover if a number is a multiple of 5 is to see if it ends with 5 or 0.

The sum of the digits of a multiple of 3 is also a multiple of 3. For example, we know that 78126 is a multiple of 3 because

$$7+8+1+2+6 = 24 = 3\times 8.$$

Work out a test which will tell you whether a number is a common multiple of 3 and 5. Which number must also be a factor of such a common multiple? Which of the following numbers is a common multiple of 3 and 5?

(*a*) 3065; (*b*) 8065; (*c*) 4065.

3. TRIANGLE NUMBERS

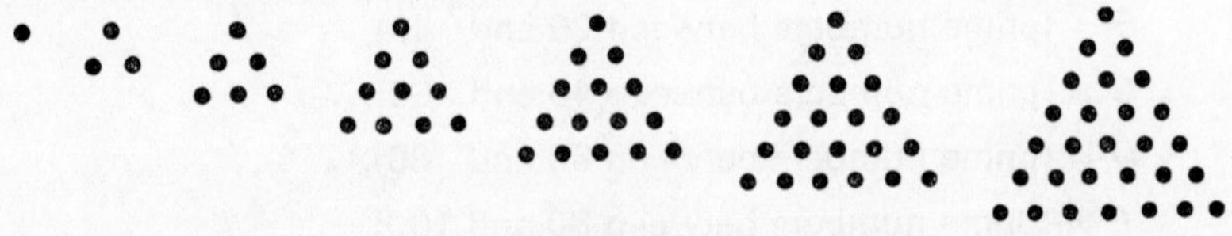

Fig. 1

These patterns of dots give the *triangle numbers*. What triangle numbers are shown? How is each triangle made from the one before? What are the next three triangle numbers? What is the difference between the thirteenth and the twelfth triangle numbers?

We can draw the same triangles in another way. Copy the triangles shown below, then draw two more to show the seventh and eighth triangle numbers.

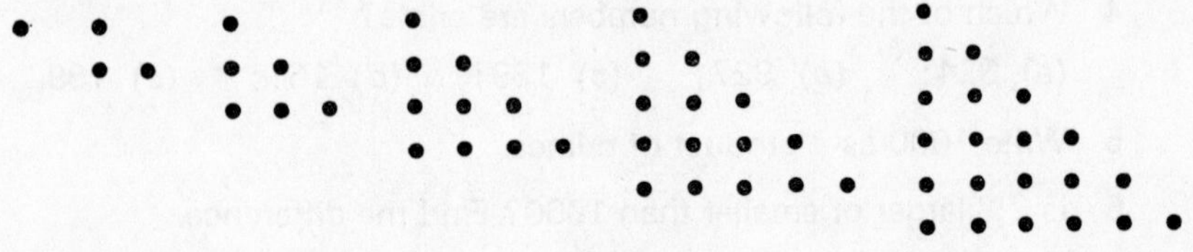

Fig. 2

Look at the patterns in Figure 3. Using both black and red dots we have put two triangle numbers together in each pattern. Does this help you to see a connection between triangle numbers and square numbers?

8 The number that is a common multiple of 3 and 5 must end in 5 or 0 and the sum of its digits must be a multiple of 3. 15. (*c*).

3. TRIANGLE NUMBERS

This section deals with sequences of numbers. The text attempts to suggest the connection between one triangle number and the next without actually stating it. Many pupils like working with sequences but become discouraged when they cannot find the connection between the terms. If the text does not give a suffient lead in this matter, it might be worth while to allow pupils to make up sequences given the first-order differences. They could work in pairs, making up and unravelling each other's sequences. Second-order differences may very well be suggested.

Triangle numbers	1		3		6		10		15		21...	
First-order differences		2		3		4		5		6		...

The connection between consecutive triangle numbers a and b where a is the nth triangle number is $b = a+n+1$.

As the nth triangle number is given by $\frac{1}{2}n(n+1)$, n can be given as $\frac{1}{2}\{^-1 + \sqrt{(1+8a)}\}$. Substituting, this gives

$$b = \tfrac{1}{2}[2a+1+\sqrt{(8a+1)}].$$

Exercise D

1 (*a*) 9; (*b*) 16; (*c*) 25; (*d*) 121.

3.1 Triangle numbers and rectangle numbers

12 is double the third triangle number.

The 5th triangle number is $\frac{5\times 6}{2}=15$.

The 6th triangle number is $\frac{6\times 7}{2}=21$.

The 7th triangle number is $\frac{7\times 8}{2}=28$.

The 10th triangle number is $\frac{10\times 11}{2}=55$.

The 11th triangle number is $\frac{11\times 12}{2}=66$.

Thus the nth triangle number is $\frac{n(n+1)}{2}$.

The only triangle number which is also a prime number is 3. The reason is that the triangle numbers are formed by half the product of two consecutive numbers and thus the result must be a rectangle number. The only exceptions are when one of the two original numbers is 2: this gives the triangle numbers 1 and 3 and of these, only 3 is prime.

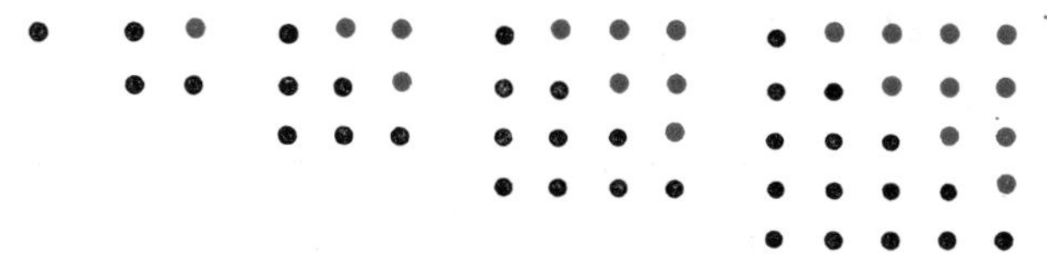

Fig. 3

Copy Figure 3 and add three more such patterns.

Exercise D

1 What is the sum of:

(*a*) the 2nd and 3rd triangle numbers;

(*b*) the 3rd and 4th triangle numbers;

(*c*) the 4th and 5th triangle numbers;

(*d*) the 10th and 11th triangle numbers?

3.1 Triangle numbers and rectangle numbers

In Figure 4 we have a slightly different pattern of red and black dots. This time we have shown each triangle number twice on the same pattern so that, apart from the first, they make rectangular patterns.

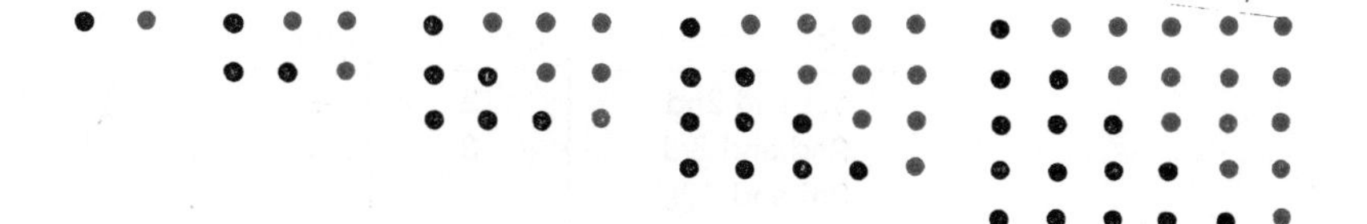

Fig. 4

Which rectangle number is double the third triangle number? The 4th triangle number is $\frac{1}{2}(4 \times 5) = 10$. Write down the 5th, 6th, 7th, 10th and 11th triangle numbers using this method. Do your answers agree with your earlier results?

Can you show that only one of the triangle numbers is a prime number? Which one is it? (If you find this rather hard—have a look at Figure 4.)

4. ALL SORTS OF NUMBERS

Exercise E

In Questions 1 and 2 the letters P, R, S and T stand for the sets

$P = \{$prime numbers less than 40$\}$,
$R = \{$rectangle numbers less than 40$\}$,
$S = \{$square numbers less than 40$\}$,
$T = \{$triangle numbers less than 40$\}$.

1 To which of the sets P, R, S, T do these numbers belong:

(*a*) 1; (*b*) 5; (*c*) 6; (*d*) 9; (*e*) 17; (*f*) 21; (*g*) 36?

2 (*a*) Write down the members of P, R, S and T.

(*b*) What can you say about:

(i) $P \cap R$; (ii) $P \cap S$; (iii) $P \cap T$; (iv) $R \cap S$?

(*c*) How many members have:

(i) P; (ii) T; (iii) $P \cap T$?

3 In this question the letters P, R, S and T stand for the sets:

$P = \{$prime numbers less than 100$\}$,
$R = \{$rectangle numbers less than 100$\}$,
$S = \{$square numbers less than 100$\}$,
$T = \{$triangle numbers less than 100$\}$.

List the members of:

(*a*) $S \cap R$; (*b*) $S \cap T$; (*c*) $P \cap T$; (*d*) $R \cap T$.

(It will help if you write out the members of T first.)

4 Copy and complete the following table:

Triangle numbers	Difference	Sum
1st and 2nd	2	4
2nd and 3rd	3	
3rd and 4th		
6th and 7th		
10th and 11th		

Can you find a connection between the 2nd and 3rd columns?

5 Cube Numbers. How many cube blocks of side 1 cm can be fitted into a cubical box whose inner edges are all of length:

(*a*) 1 cm; (*b*) 2 cm; (*c*) 3 cm?

The answers are all members of the set of cube numbers. Can you find three more such numbers to add to the set? (Remember that we write 2 cubed as 2^3.)

4. ALL SORTS OF NUMBERS

Exercise E

1 (*a*) $1 \in S$, $1 \in T$; (*b*) $5 \in P$; (*c*) $6 \in R$, $6 \in T$;
(*d*) $9 \in R$, $8 \in S$; (*e*) $17 \in P$;
(*f*) $21 \in R$, $21 \in T$; (*g*) $36 \in R$, $36 \in S$, $36 \in T$.

2 (*a*) $P = \{2, 3, 5, 7, 11, 13, 17, 19, 23, 29, 31, 37\}$,
$R = \{4, 6, 8, 9, 10, 12, 14, 15, 16, 18, 20, 21, 22, 24, 25, 26, 27, 28, 30, 32, 33, 34, 35, 36, 38, 39\}$,
$S = \{1, 4, 9, 16, 25, 36\}$,
$T = \{1, 3, 6, 10, 15, 21, 28, 36\}$;
(*b*) (i) $P \cap R = \varnothing$, (ii) $P \cap S = \varnothing$, (iii) $P \cap T = \{3\}$,
(iv) $R \cap S = \{4, 9, 16, 25, 36\}$.

This part of the question affords the opportunity to give the notation for the empty set, $\varnothing$.

(*c*) (i) 12, (ii) 8, (iii) 1.

3 (*a*) $S \cap R = \{4, 9, 16, 25, 36, 49, 64, 81\}$;
(*b*) $S \cap T = \{1, 36\}$;
(*c*) $P \cap T = \{3\}$;
(*d*) $R \cap T = \{6, 10, 15, 21, 28, 36, 45, 55, 66, 78, 91\}$.

4

2	4
3	9
4	16
7	49
11	121

Sum = (difference)^2.

5 (*a*) 1; (*b*) 8; (*c*) 27. $4^3 = 64$, $5^3 = 125$, $6^3 = 216$.

6 The first-order differences for the pyramid numbers are the square numbers, for each layer will be a square one unit larger than the one above. Using this idea and writing first-order differences first we have:

First-order differences		4	9	16	25	36	...
Pyramid numbers	1	5	14	30	55	91...	

6 Pyramid Numbers. A pyramid can be made by placing 9 tennis balls in a square, then adding another layer with 4 balls and a top layer of 1 ball. How many balls are needed altogether?

The answer to this is a member of the set of pyramid numbers. Add three more pyramid numbers to those given:

1, 5, 14, ...

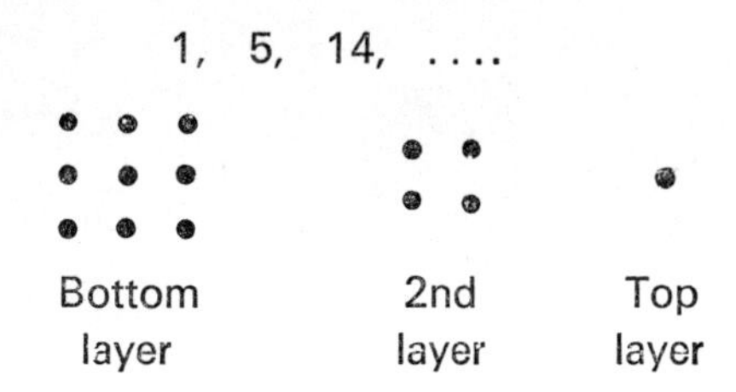

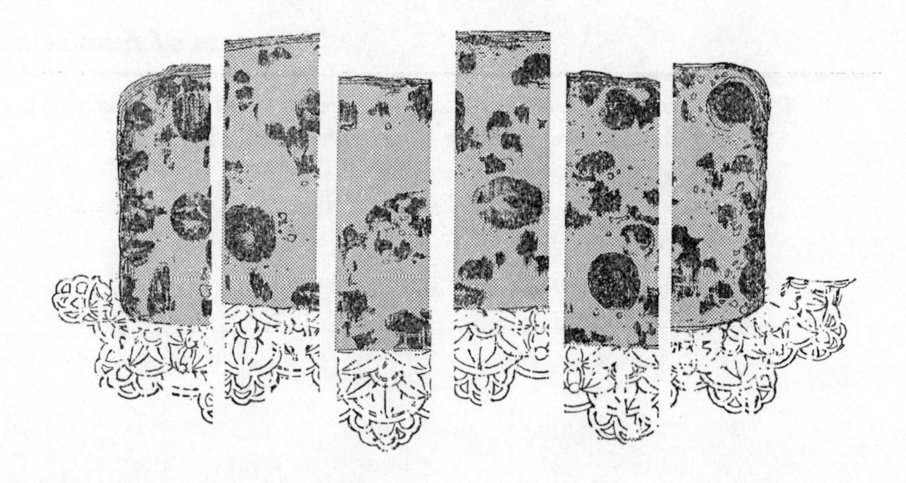

9. Two ways of looking at division

1. REPEATED SUBTRACTION

(*a*) Some people are queueing up for sandwiches, each made from two slices of bread. There are 20 slices, how many are left after:

1 person has taken a sandwich;

2 people have taken a sandwich each;

3 people have taken a sandwich each?

This is a process of 'repeated subtraction'. It can be continued until there is either one or no slice left. How many people can get a sandwich if there were 17 slices to start with? How many people if there were 117 slices to start with; or 1117?

All those subtractions would have wasted time so you probably used division. In this case division is a quick way of carrying out repeated subtraction, (see Chapter 4, Section 1.2). (For some computers, repeated subtraction is the only way of carrying out division.)

What is left can be stated either as 1 slice, or, as a sandwich is made from 2 slices, as a $\frac{1}{2}$ sandwich.

(*b*) Draw some diagrams to show how 20 bars of chocolate can be arranged in stacks having: (i) 3 bars in each, (ii) 4 bars in each, (iii) 7 bars in each. How many stacks are there in each case? What fraction of a stack is left over?

(*c*) Some children are counting a great pile of tenpenny pieces.

9. Two ways of looking at division

A formal approach to fractions starts by defining the fraction b/a to be the unique solution of the equation $ax = b (a \neq 0)$ where a and b are integers. All the familiar properties of fractions now follow, e.g.

$$\frac{b}{a}+\frac{d}{c}=\frac{d}{c}+\frac{b}{a} \quad \text{or} \quad \frac{b}{a}+\frac{d}{c}=\frac{bc+ad}{ac}.$$

There remains the question 'Does the solution to the equation $ax = b$ exist?' (This is like the question 'Does $\sqrt{-1}$ exist?'.) It does if we can construct it from numbers that we know exist. To construct a solution to $ax = b$ we consider the ordered pair (b, a) with a and b integers and $a \neq 0$. We now define equivalence, addition and multiplication of ordered pairs in such a way as to lead to the equivalence, addition and multiplication of fractions. So, for example:

$$(b, a) = (d, c) \quad \text{if and only if} \quad bc = da,$$

$$(b, a) + (d, c) = (bc + da, ac),$$

$$(b, a) \times (d, c) = (bd, ac).$$

It turns out that (b, a) is the solution to the equation $ax = b$. The details of the argument can be found in such books as *Elements of Algebra,* by H. Levi and *The Number System,* by H. A. Thurston.

This is one way a formal study of rational numbers would start and it is too complicated for an introduction at this level.

We have fallen back upon the more practical method involving illustrations and, for this chapter, a familiarity with the techniques of division (though not considering it as an inverse of multiplication in any formal sense).

But, though pupils are familiar with the actual process of division, they are not usually familiar with the various interpretations of this operation. The purpose of this chapter is to bring these ideas forward so that pupils may understand what they are doing when they divide, and also why it is that numbers are written one over the other, in the form of a fraction, in order to indicate that one is to be divided by the other.

Division

In Chapter 6, we establish first that a fraction is a number which can represent part of a whole and then that it is a number of the form b/a where a and b are integers and $a \neq 0$. The next step is to show why we use a fraction to indicate a process of division. For example, we write $\frac{7}{2}$ to indicate $7 \div 2$. What is this to do with parts of a whole?

In dealing with objects in practical situations, 'division' is the name given to two quite distinct processes (though, of course, the mathematical signs and manipulative techniques are the same in both cases). It can be a short method of finding the answer to a process of *repeated subtraction*. 'How many groups of 12 consecutive notes are there on a piano keyboard of 88 notes?'

It can also be considered as the inverse of multiplication. Multiplication tells us the total if the number of objects in a group is multiplied by the number of groups. Division, in this sense, tells us the number of objects in a group if the total is divided by the number of groups. 'He divided his fortune equally between his three daughters.'

Repeated subtraction. The question, 'How many groups of three people can we form if there are 16 people altogether?', can be answered on paper by dividing 16 by 3. We can obtain the answer in practice by 'subtracting' groups of three from the total number until only one person is left. The division process gives us $16 \div 3 = 5\frac{1}{3}$. The practical process of repeated subtraction gives us five groups of three and one person, i.e. one-third of a group left over. Now we know from Chapter 6 that $5\frac{1}{3}$ is the same as $\frac{16}{3}$ and so it seems that the answer to $16 \div 3$ can be written as the fraction $\frac{16}{3}$. For this reason we tend to use $\frac{16}{3}$ as an alternative way of writing $16 \div 3$.

It is this approach to division that is the basis of a technique for long division and it is this method that is used in the calculating machine or computer.

Inverse multiplication. If sixteen objects are divided into three equal groups, then, from the process of division we know that each group must contain five and a third objects. This is confirmed if we try to draw or visualize this situation. Using the above argument we again find that we can write $16 \div 3 = \frac{16}{3}$.

They are doing it by arranging them in 'pound' heaps. Is this repeated subtraction? If you thought that there were 1041 tenpenny pieces altogether, how many heaps would you expect? What fraction of a heap is left?

(*d*) Some extra people joined the queue for sandwiches when there were 117 slices. It was decided to serve $\frac{1}{2}$ sandwiches. How many half sandwiches were available?

117 halves can be written $\frac{117}{2}$ or, from Chapter 6, as $58\frac{1}{2}$. But, in (*a*), we did the division $117 \div 2$ to obtain the answer $58\frac{1}{2}$.

As $\frac{117}{2}$ is a way of writing the answer to $117 \div 2$, it has become a shorthand for division and, for example, $\frac{13}{2}$ may be read '13 divided by 2'.

2. FAIR SHARES

A mother has a family of four children who always insist on their 'fair' share of food! One day she sends them to a nearby shop with Order I. Instead they bring back the things in Order II.

Order I	*Order II*
1 large cream cake	1 large cream cake
2 scones	2 scones
4 doughnuts	3 doughnuts
4 currant buns	5 currant buns
4 tomato sandwiches	4 sandwiches (1 each of cheese, ham, tomato and beef)

Her original order would have been quite easy to divide equally among the four children. How would you have done this? The second order is not sc easy, but the mother managed to do it by cutting carefully like this:

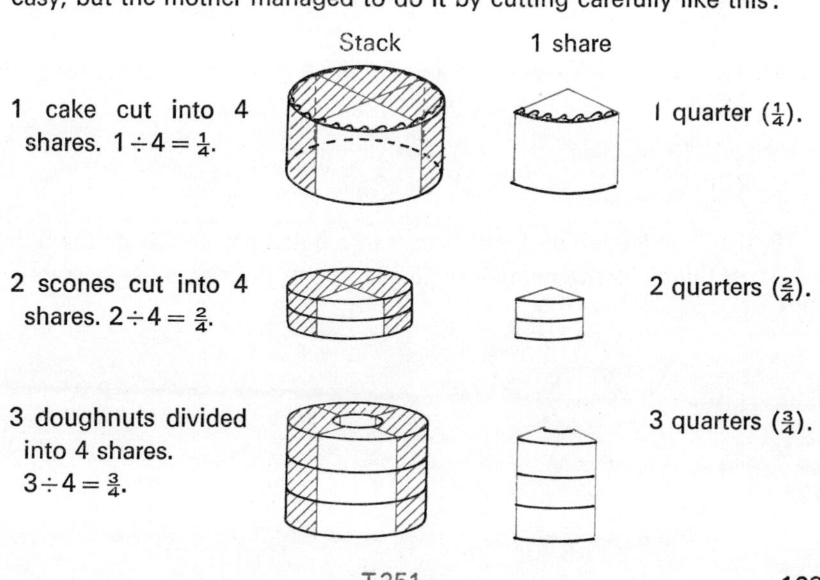

4 sandwiches divided into 4 shares. $4 \div 4 = \frac{4}{4}$.

4 quarters ($\frac{4}{4}$). (Notice how 'fair' the mother was here.)

5 currant buns divided into 4 shares. $5 \div 4 = \frac{5}{4}$.

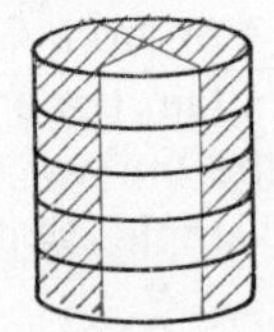

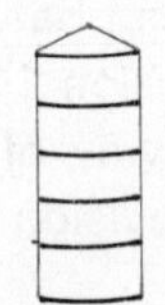

5 quarters ($\frac{5}{4}$).

Whatever number of cakes or sandwiches the mother had, she could always obtain 4 fair shares by this method. Any number can be divided by 4.

Suppose, however, that there had been six children, not four. Draw a picture to show how she would have divided the four sandwiches into 6 equal shares.

To find the number of sweets that each would get if there were 30 altogether, you would again do a division 'sum'. This time we are using division as a method for finding the answer when splitting some collection of objects into equal groups. It is the opposite of multiplication which gives the answer when groups of the same sizes are joined together.

Is it possible to divide by *any* number? If not, what is the exceptional number?

Exercise A

1 The following sums of money are to be divided equally between two charities. How much would each charity receive?

(*a*) 13 pennies; (*b*) 101 pennies; (*c*) £48;
(*d*) £7; (*e*) £12345.

2 Each collection of things is split into equal shares. Divide the number of things by the number of shares. Is this the size of each share?

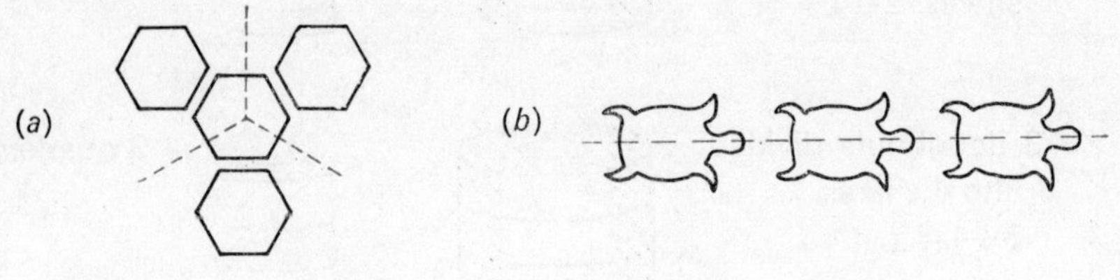

4 things in 3 shares   3 things in 2 shares

1. REPEATED SUBTRACTION

We hope that pupils will use their imagination and picture people taking two slices of bread from the pile for each sandwich and continuing to do this until there are less than two slices left. Similarly in (*b*) (i), we should expect them to draw a stack of three, then another stack of three and then another, until there are less than three bars left. Each time they draw a stack of three they should consider that three bars have been subtracted from the total available. In (*c*) the tenpenny pieces are being taken from the central pile ten at a time. This again is repeated subtraction.

2. FAIR SHARES

As before, we hope that pupils will imagine the situations and realize the difficulties involved in dividing quantities into equal parts.

It is considered impossible to divide by zero. It is better to state that the answer to such a division is undefined rather than that the answer is infinity, because this introduces difficulties which cannot satisfactorily be dealt with at this stage. Interpreting division by zero in the context of a computer, it amounts to repeated subtraction of nothing and this would continue indefinitely.

Exercise A

1 (*a*) $6\frac{1}{2}$p; (*b*) $50\frac{1}{2}$p; (*c*) £24;
(*d*) £$3\frac{1}{2}$; (*e*) £$6172\frac{1}{2}$.

2 The divisions required are simple. The question is intended to give a visual check. However, some pupils may derive benefit from closely relating the figures to the numbers given.

(*a*) $1\frac{1}{3}$; (*b*) $1\frac{1}{2}$; (*c*) $6\frac{1}{2}$; (*d*) $3\frac{1}{3}$.

3 This question may also be expressed as a division of two numbers. The pupils will probably find it easiest to look at the left-hand part in order to express its length in terms of a whole unit.

(*a*) $\frac{2}{3}$ units; (*b*) $1\frac{2}{3}$ units;
(*c*) $\frac{2}{5}$ units; (*d*) $\frac{3}{4}$ units.

4 (*a*) $8 \div 2 = 4$; (*b*) $13 \div 3 = 4\frac{1}{3}$;
(*c*) $5 \div 4 = 1\frac{1}{4}$; (*d*) $7 \div 3 = 2\frac{1}{3}$;
(*e*) $11 \div 3 = 3\frac{2}{3}$; (*f*) $5 \div 10 = \frac{1}{2}$.

Warning: these examples have been chosen carefully. The splitting is shown symmetrically so that the groups are obviously equal.

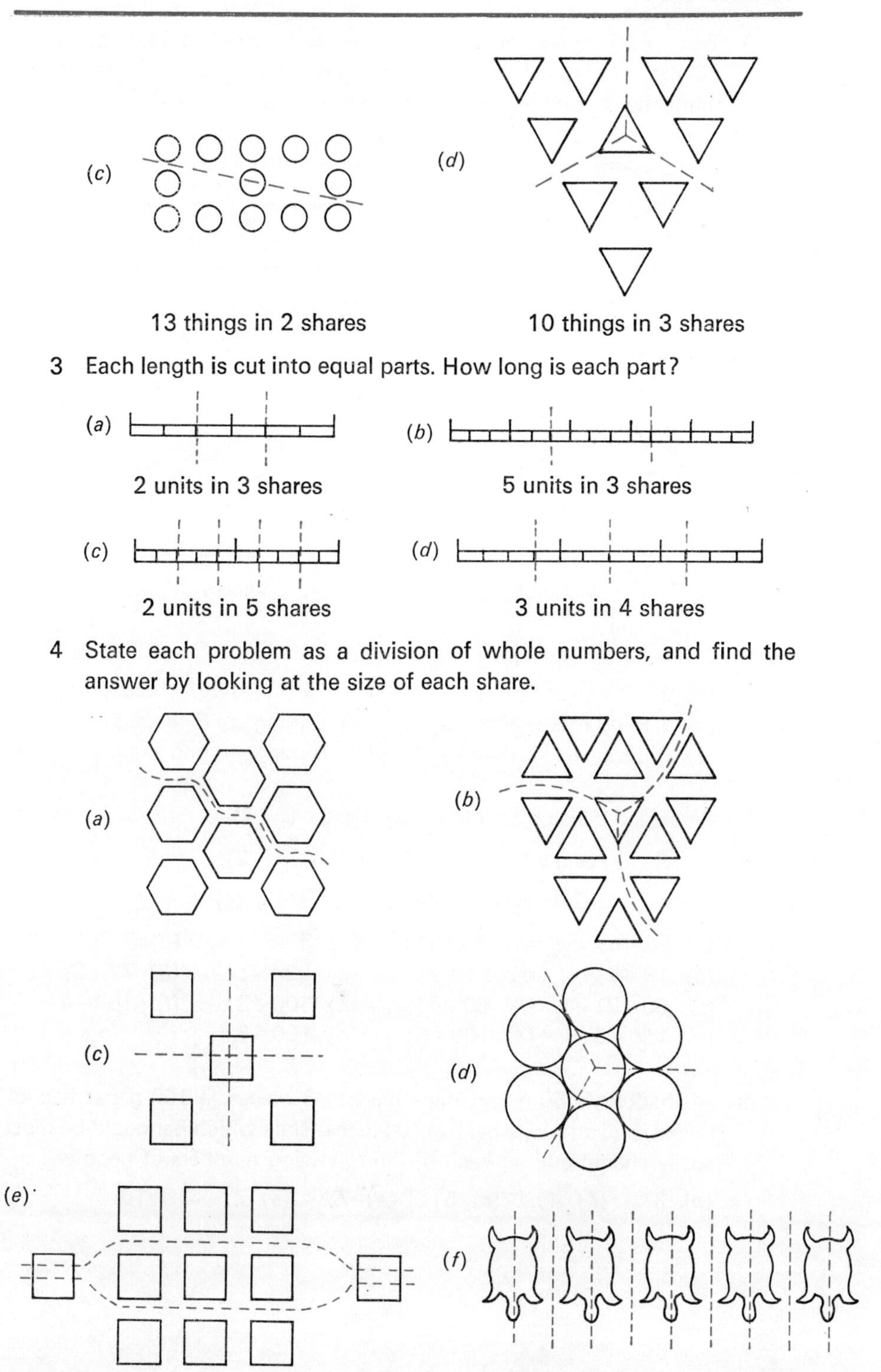

(c) 13 things in 2 shares

(d) 10 things in 3 shares

3 Each length is cut into equal parts. How long is each part?

(a) 2 units in 3 shares

(b) 5 units in 3 shares

(c) 2 units in 5 shares

(d) 3 units in 4 shares

4 State each problem as a division of whole numbers, and find the answer by looking at the size of each share.

(a)

(b)

(c)

(d)

(e)

(f)

5 Trace each group of things. Divide each diagram into the stated number of equal shares, and shade one of these shares. How many things have you shaded?

(*a*) 2 shares

(*b*) 4 shares

(*c*)

4 shares

(*d*)

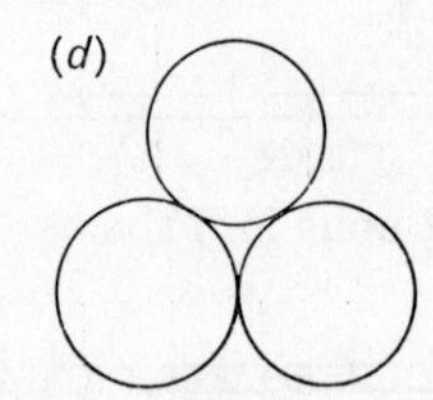

6 shares

6 Divide to find the size of each share:

(*a*) 14 things in 2 shares; (*b*) 15 things in 4 shares;
(*c*) 1 thing in 2 shares; (*d*) 6 things in 6 shares;
(*e*) 6 things in 12 shares; (*f*) 6 things in one share.

7 Give each division as a fraction (e.g. $5 \div 9 = \frac{5}{9}$),
as a whole number (e.g. $18 \div 9 = 2$),
or, as a mixed number (e.g. $23 \div 9 = 2\frac{5}{9}$).

(*a*) $10 \div 5$; (*b*) $5 \div 10$; (*c*) $5 \div 5$; (*d*) $7 \div 5$;
(*e*) $18 \div 7$; (*f*) $21 \div 7$; (*g*) $22 \div 7$; (*h*) $72 \div 2$;
(*i*) $50 \div 7$; (*j*) $50 \div 4$; (*k*) $100 \div 3$; (*l*) $100 \div 4$;
(*m*) $100 \div 5$; (*n*) $100 \div 6$; (*o*) $100 \div 7$.

8 A Chadbury's 50 g chocolate bar has 8 pieces, a 100 g bar has 21 pieces, and an Airlite bar has 10 pieces. State which bar could be most easily shared among each of the following numbers of people:

(*a*) 3; (*b*) 4; (*c*) 5; (*d*) 7; (*e*) 2; (*f*) 16.

5

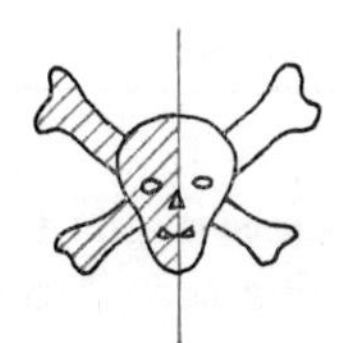

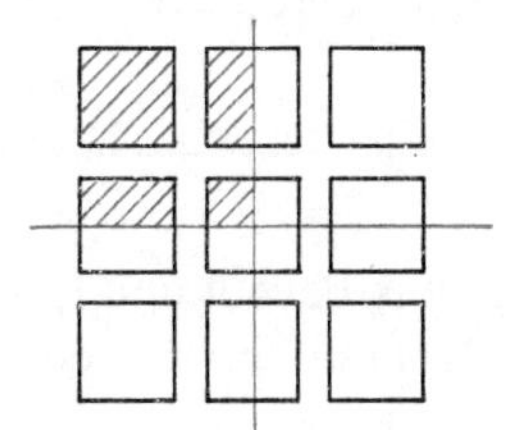

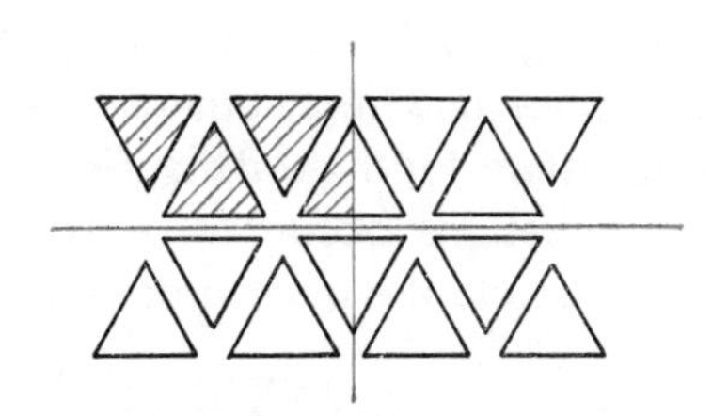

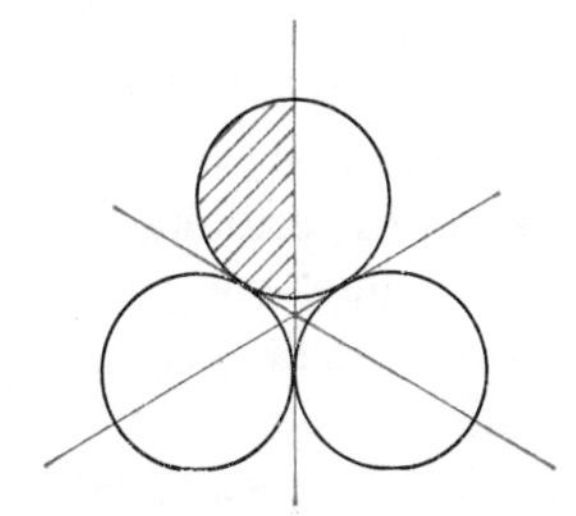

Notice that there are several acceptable answers to (*b*) and (*c*) and three for (*d*).

6 (*a*) 7; (*b*) $3\frac{3}{4}$;
(*c*) $\frac{1}{2}$; (*d*) 1;
(*e*) $\frac{1}{2}$; (*f*) 6.

7 (*a*) 2; (*b*) $\frac{1}{2}$; (*c*) 1; (*d*) $\frac{7}{5}=1\frac{2}{5}$;
(*e*) $\frac{18}{7}=2\frac{4}{7}$; (*f*) 3; (*g*) $\frac{22}{7}=3\frac{1}{7}$; (*h*) 36;
(*i*) $\frac{50}{7}=7\frac{1}{7}$; (*j*) $\frac{50}{4}=12\frac{1}{2}$; (*k*) $\frac{100}{3}=33\frac{1}{3}$; (*l*) 25;
(*m*) 20; (*n*) $\frac{100}{6}=16\frac{2}{3}$; (*o*) $\frac{100}{7}=14\frac{2}{7}$.

8 (*a*) 100 g bar; 21 is divisible exactly by 3.
(*b*) 50 g bar; 8 is divisible exactly by 4.
(*c*) Airlite bar; 10 is divisible exactly by 5.
(*d*) 100 g bar; 21 is divisible exactly by 7.
(*e*) 50 g bar or Airlite; 8 and 10 are both divisible exactly by 2.
(*f*) 50 g bar; $8\div16=\frac{1}{2}$ which gives by far the easiest fraction of a piece.

T

10. Polyhedra

The authors have found it rewarding to allow pupils to spend some considerable time on the content of this chapter. Pupils generally enjoy constructing polyhedra. There are, however, good mathematical reasons for studying this topic. Some of the work already covered on polygons can be revised in an interesting context. The desire to obtain satisfactory results encourages pupils to regard their drawings with a critical eye and consequently to strive for greater accuracy than they might otherwise achieve. There is an opportunity to develop a feeling for three-dimensional work. Also, some interesting mathematical situations occur; for example, the number of possible nets for a given solid, and the question of impossible polyhedra.

Apart from occasions when the construction of a net is an exercise in itself, as it will be in the early stages, it is useful to have a means of mass-producing a net. It is worthwhile taking the trouble to duplicate a number of nets which can then be used as templates for pricking through onto cartridge paper. Each new net should always be pricked through from a 'master' for otherwise inaccuracies can rapidly creep in.

Size is also important; if a net is made too small then the construction can be tricky. Equilateral triangles of side 4–5 cm give a good result for deltahedra; pentagons of side 4–5 cm give a good result for the dodecahedron.

Squared paper and isometric paper are extremely useful for sketching polyominoes and diamond-patterns. Isometric paper with equilateral triangles of side 2·5 cm is an appropriate size; it is sometimes difficult to obtain this commercially and worth while to produce one's own.

1. NETS

The aim of this and succeeding sections is not only to develop competence at producing a particular polyhedron from a given net but, what is more important, to enable the pupil to design a net for a *given* polyhedron. Much more will have been achieved if this second objective is reached, even if only in simple cases.

Questions (*a*), (*b*), (*c*) in the text are more suitable for a class project or group work than for individual work. Each 'new' shape can be displayed as it is found. At first there may be repetitions which pass unnoticed by the class, as some pupils have difficulty in recognizing a shape when it is turned over or turned round, but gradually repetitions will be eliminated. The shapes can then be sorted into two sets: those that are nets for a cube and those that are not.

(*a*) The first shape in Figure 2 is the popular net for the cube. The second shape will not form a cube but the third will.

(*b*) There are thirty-five different ways, excluding mirror images, in which six squares can be joined edge to edge. These shapes are called hexominoes and eleven of them are nets for a cube. The hexominoes are shown in Figure A with the eleven that form a cube in Figure A (i).

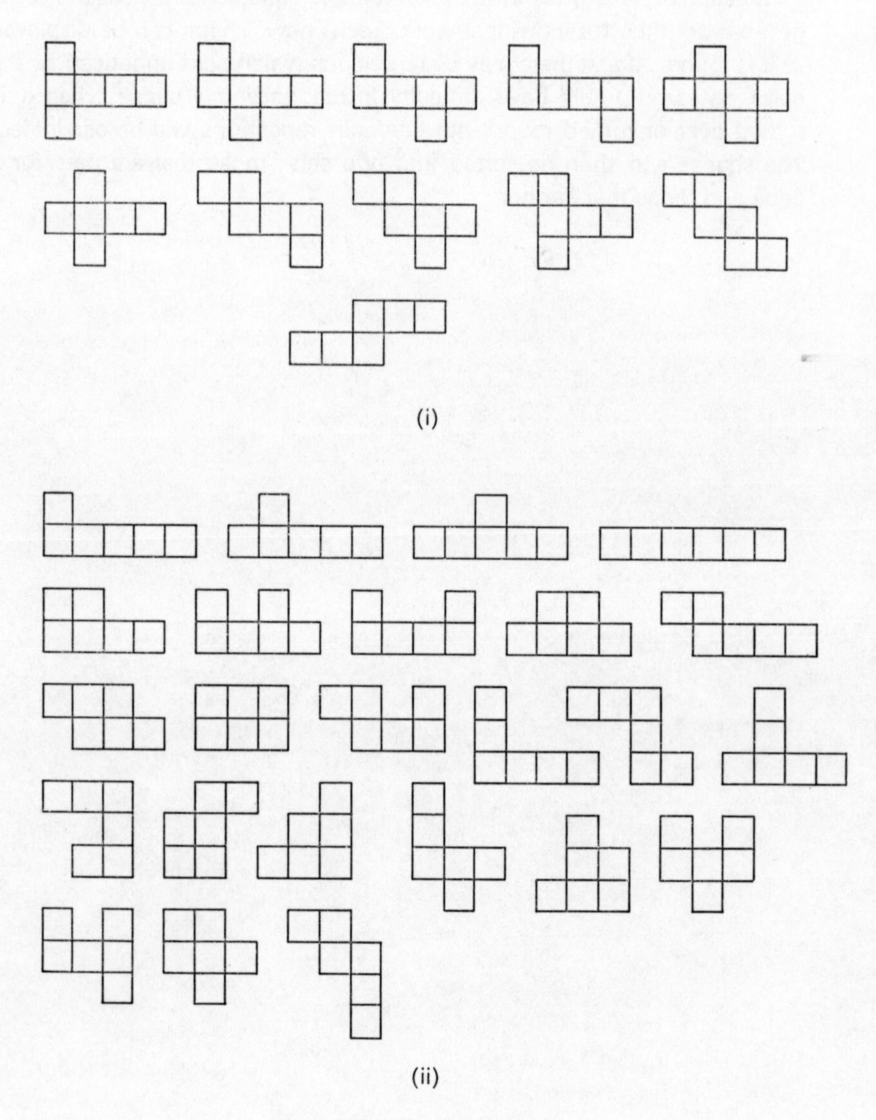

Fig. A. Thirty-five hexominoes.

10. Polyhedra

1. NETS

To make a hollow cube, Figure 1, six equal squares are needed. We could cut out all six and then join them with adhesive tape but, as you can see from any cardboard box, this is not necessary. We could draw six equal squares on paper as in Figure 2, cut round the outside and then *fold* them up into a cube.

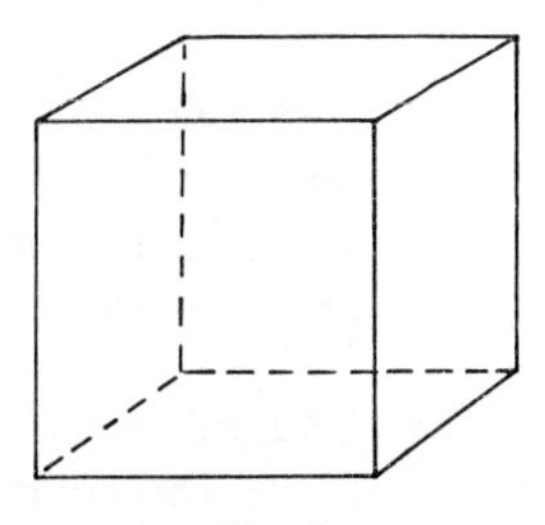

Fig. 1

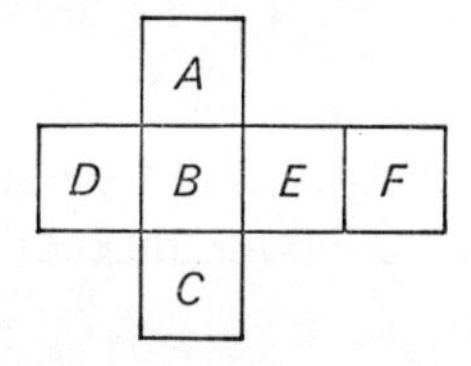

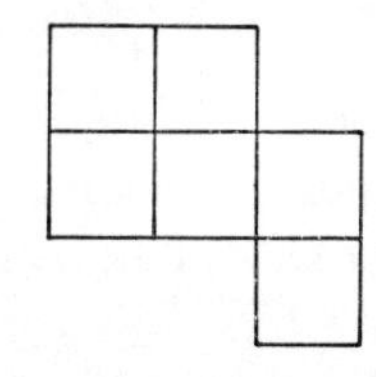

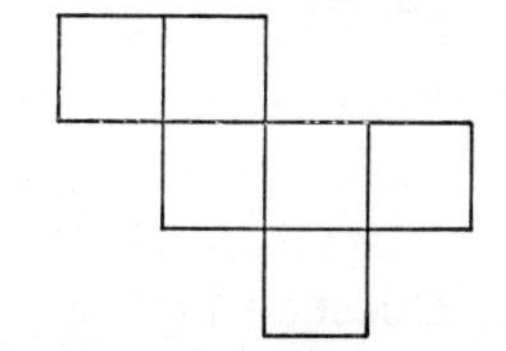

Fig. 2

Six squares can be joined edge to edge in several ways. Figure 2 shows three examples. In the first example, squares *A, D, E, C* could be folded up to form the side faces of a cube with *B* the bottom face and *F* the top face.

(*a*) Can either of the other shapes be folded to form a cube? If you are not sure, copy these shapes on to squared paper. Then cut them out and fold them.

(*b*) On squared paper, draw as many other shapes as you can, made with six squares in this way. Try to decide which of them could be folded to form a cube. Check your results by cutting and folding. How many shapes can you find? Has your neighbour found any shapes which are different from yours?

(*c*) Each of the shapes which can be folded to form a cube is called a *net* of the cube. Which of your shapes are *obviously* not nets of the cube? Why?

Solid figures which have plane faces are called *polyhedra* which means 'having many faces'. (Singular–polyhedron.) In this section, we shall not make the solid figures but only their surfaces.

Exercise A

1 Draw as many shapes as you can by joining five equal squares edge to edge.

Mark those which could be folded to form the bottom and sides of a cubical box without a lid. Exchange books with a neighbour. Do you agree with your neighbour's results?

2 Can you make a rectangular box (or 'cuboid'), using some adhesive tape and two rectangular cards 16 cm by 8 cm, two 8 cm by 6 cm, and two 16 cm by 6 cm? If so, sketch a net for the box. Show the measurements of the cards on your sketch.

3 You are given two rectangular cards 6 cm by 3 cm and two 6 cm by 2 cm. You need two more cards to make a cuboid. What is the size of the missing cards? Draw an accurate net for the cuboid on squared paper.

4 Draw several nets for a rectangular box without a lid given the following measurements: 3 cm long, 2 cm wide and 1 cm high.

Are there as many possible nets as you found for the cubical box in Question 1?

D5 Cubes, cuboids and the other figures in Figure 3 are all examples of a special type of polyhedron called a *prism*. They all have two ends (those marked in the figure) which are parallel and which are the same shape and size. The faces joining the ends must be parallelograms.

Examples of prisms are an unsharpened pencil (see Figure 3), a ruler, a fiftypenny piece.

Give some more examples. Is a cube a prism?

The ends of the unsharpened pencil in Figure 3 are regular polygons. Name them.

D6 A prism with circular ends is called a *cylinder*. Are these cylinders: a broom handle, a cocoa tin, a penny?

(*c*) Some pupils will quickly see that shapes containing four squares surrounding a point such as the second shape in Figure 2 cannot possibly form a cube; others will need to cut and fold the shapes. It is also obvious to some pupils that shapes with five or six squares in a line cannot form a cube. Others may see that if four squares are in line, then the remaining two squares must be placed on opposite sides of the 'line'.

Exercise A

1 The twelve shapes, called pentominoes, are shown in Figure B. Mirror images are again excluded. Eight shapes make a cubical box without a lid. See Figure B (i).

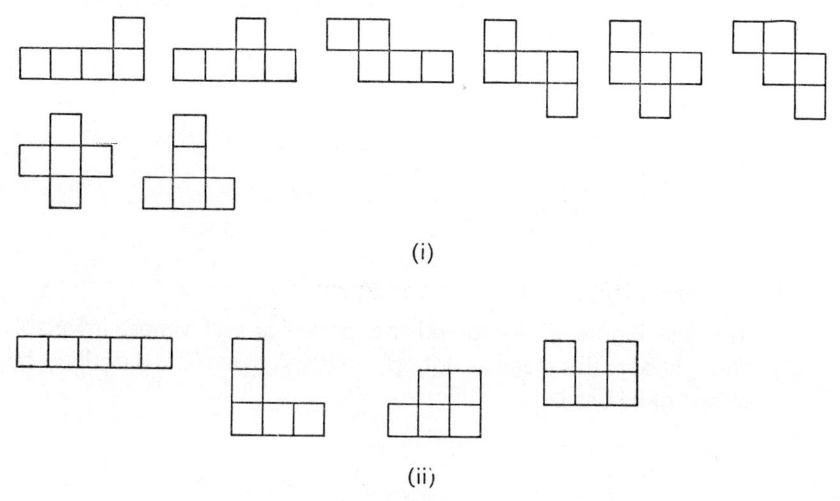

Fig. B. Twelve pentominoes.

2 One possible net is shown in Figure C.

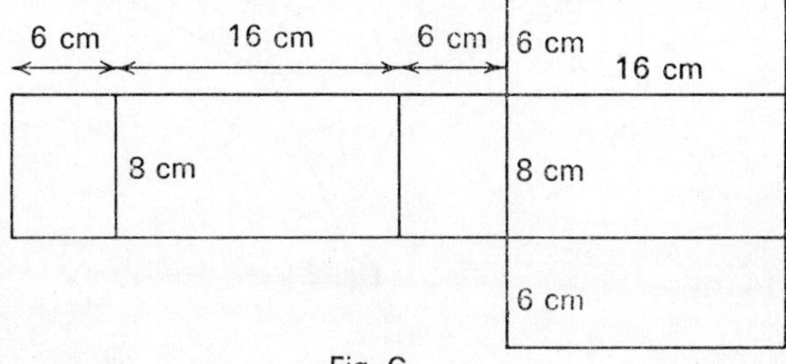

Fig. C

3 3 cm by 2 cm.

4 Yes. 8 possible nets.

5 Tent, swimming bath, drainpipe; a cube is a square-based prism. All the examples given are right prisms, that is, the lines joining corresponding vertices of the ends are perpendicular to the congruent polygons forming the 'identical ends'.

Hexagons; the unsharpened pencil is a hexagonal prism, since a prism is described by the shape of its cross-section.

7 See Figure D. For a cylinder the ends would be separate.

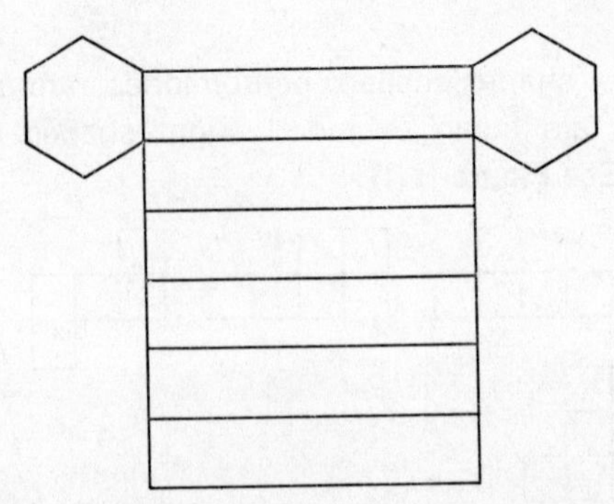

Fig. D. Net for right hexagonal prism.

8 The four triangular faces are congruent.

(*a*) See Figure E. As for prisms, pyramids will almost invariably be thought of at this stage as right pyramids. A pyramid is described by the shape of its base.

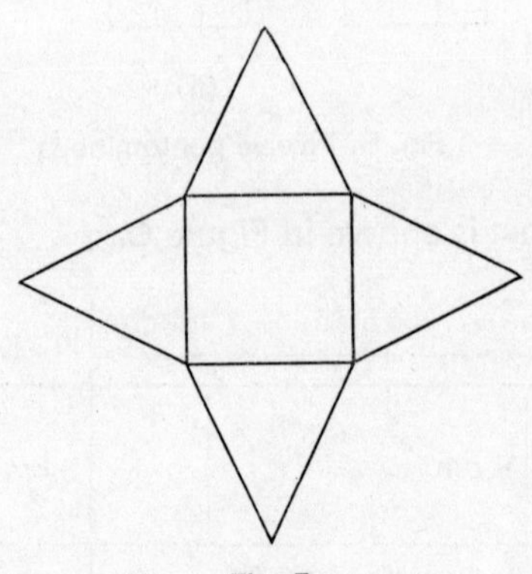

Fig. E

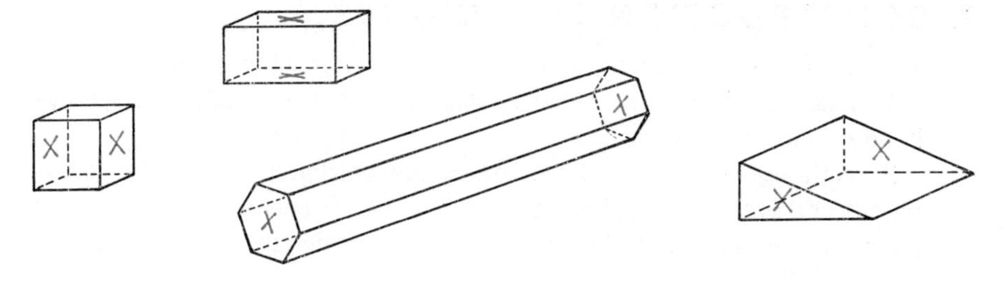

Fig. 3

7 Sketch first the net for a hexagonal prism and then the net for a cylinder. Would you have to make the ends as separate pieces?

8 Figure 4 shows a pyramid with a square base. The four triangular faces have the same shape and size and are isosceles triangles.

(*a*) Sketch a net for this pyramid.

(*b*) Suppose you wanted to construct the pyramid in Figure 4 without the base. Would it be possible to use a similar net now? If not, what different shape of net could you use? (*Hint*: draw a circle.)

(*c*) Would it be possible to include the base in the new shape of net, if you wanted it after all?

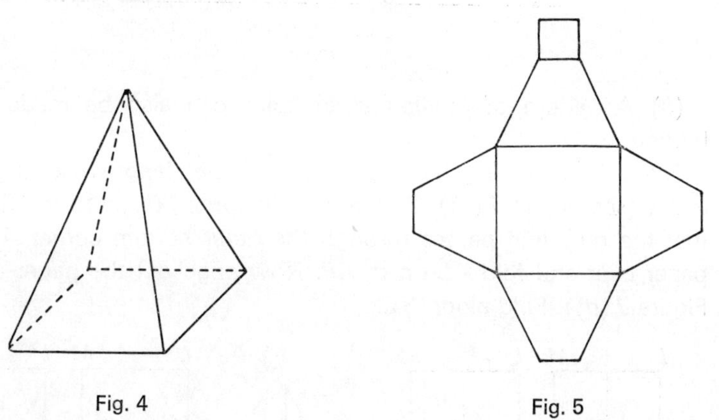

Fig. 4 Fig. 5

9 Figure 5 shows the net of a polyhedron. *Either* describe the polyhedron *or* copy the net and make the polyhedron.

10 Make an accurate model of a triangular prism, the triangular ends being equilateral.

Could you make a hexagonal prism out of a number of these triangular prisms? How many would you need?

2. CONSTRUCTION TECHNIQUES

You will find the following advice helpful when making polyhedra.

1 Accuracy in making a net is very important if a satisfactory polyhedron is to be obtained.

2 Although squares and rectangles can be drawn using ruler, compasses and protractor (or set-square), it is best to use squared paper and prick through the corners of the net on to the material being used for the model itself.

3 Triangles are best constructed with compasses.

4 (*a*) The flower pattern (see Figure 6) is really a network of equilateral triangles and is useful for making nets of polyhedra with equilateral triangular faces.

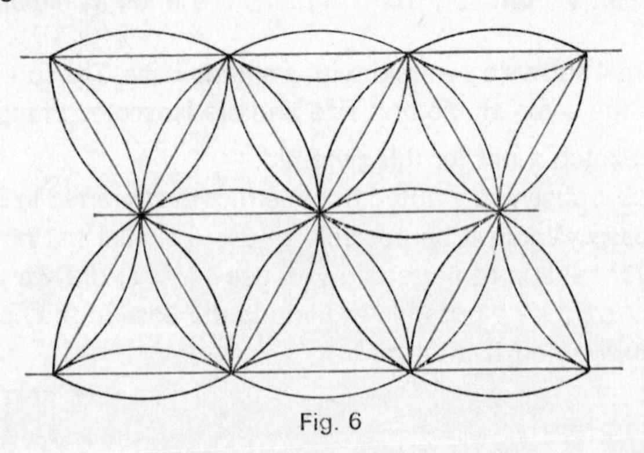

Fig. 6

(*b*) A pattern of equilateral triangles can also be made by paper folding.

Fold a *large* rectangular piece of paper in half lengthwise, and open out again (see Figure 7 (*a*)). Fold a bottom corner, *B*, on to the first fold so that the new fold passes through the other bottom corner *A*. Turn the paper over and fold *AL* on to *AP*. Now open out the paper again (see Figure 7 (*d*)). Fold along *BC*.

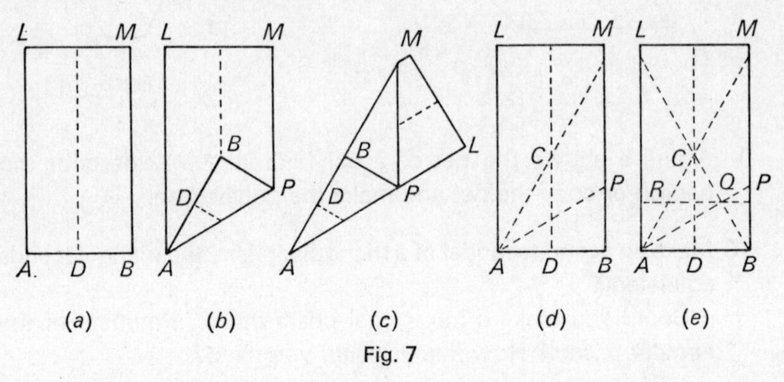

Fig. 7

(*b*) No. See Figure F. This is an important stage in the visualization of nets. So far, nets have been built up around a particular face of the polyhedron. Now they are being built up round a vertex and this is a much more useful method for general use.
(*c*) Yes. One possible position of the base is indicated by a broken line in Figure F.

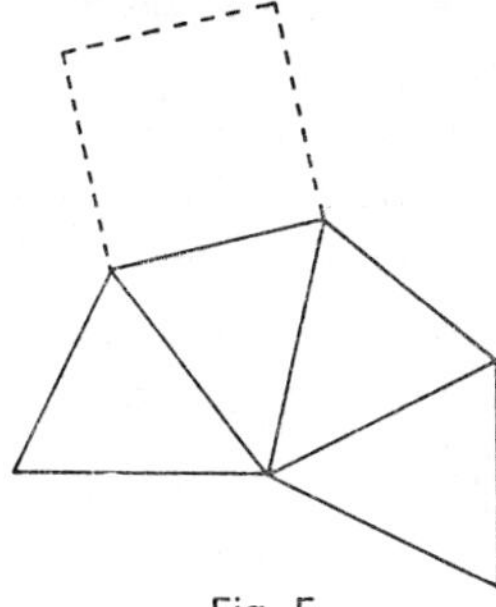

Fig. F

9 Figure 5 shows the net for a truncated square-based pyramid. (Pupils may describe it as a square pyramid with the top chopped off.) Like the cube, it is a hexahedron with twelve edges and eight vertices. Each of the four congruent faces is an isosceles trapezium.

10 Yes; six.

2. CONSTRUCTION TECHNIQUES

Pupils should be asked to make their own drawing of the flower pattern shown in Figure 6. To promote accuracy, it is a good idea to work from a base line (see Figure G).

Fig. G

Medium weight cartridge paper should be used for first attempts at constructing polyhedra; as experience is gained pupils can transfer to coloured card. Printers sometimes have offcuts of card.

A cheap round-ended pair of paper-cutting scissors costs about 5p. There should be enough for each pupil to have a pair, or for one pair to be shared between two pupils.

Two quick-drying glues are balsa-cement and Bostik I.

Tabs need only be attached to alternate edges round a net for a polyhedron. However, it is best if the last face to be secured is left free of tabs. The missing tabs must then be added to the other edges. Figure H shows a satisfactory arrangement of tabs for one of the nets for a cube. The shaded face should be secured last.

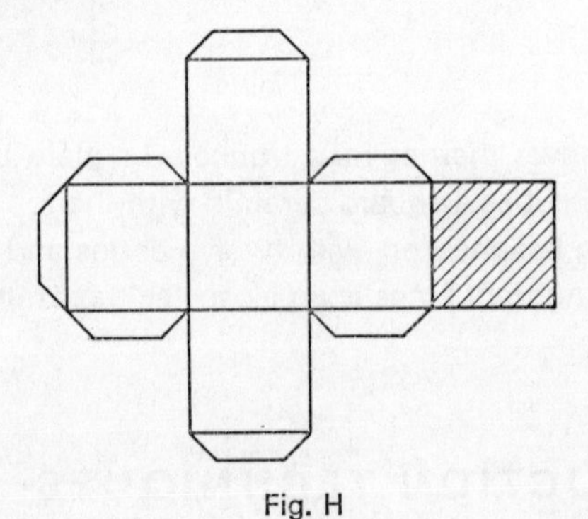

Fig. H

Models can also be made with straws joined by pipe cleaners. Alternatively, cotton can be threaded (or sucked) through each straw in turn. If a continuous thread is used, then it will have to pass through some straws twice. Problems of rigidity occur in many cases, requiring the addition of cotton diagonals.

Straws can be obtained cheaply in boxes of 500 from many multiple stores.

Painted models are very attractive. Poster paints give a matt finish and for a glossy finish the quick-drying enamel paints used for model trains are best. Cellulose enamels, which are also quick drying, can be used, though two coats may be needed to produce a good result.

Triangle ABC is equilateral.

To continue the pattern, fold triangle ABC so that D falls on C and open out again (see Figure 7 (*e*)). Now fold A onto Q and B onto R.

(*c*) The easiest way of obtaining equilateral triangles is to prick through isometric graph paper (see Figure 8). Before using this method you should be sure that you can use successfully at least one of the other two methods.

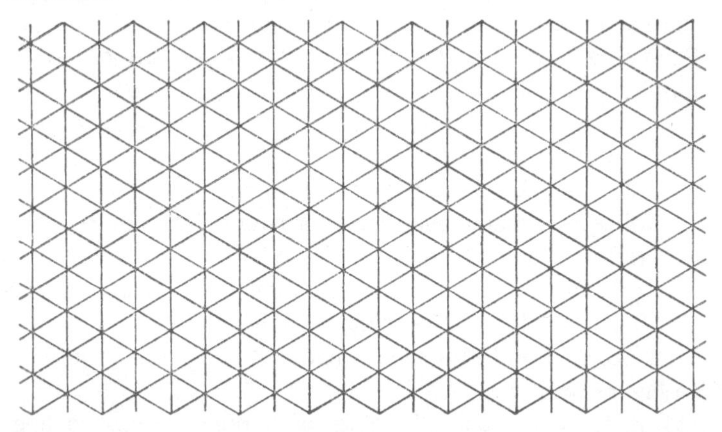

Fig. 8. Isometric graph paper

5 Coloured card or heavyweight cartridge paper are the best materials for making models. Before folding, run a compass point or other sharp instrument along the lines to be folded; in this way, a clean fold is obtained. Edges may be joined with adhesive tape, but, for a more professional result, use tabs on the edges and stick with a quick-drying glue. If you cannot decide where to put the tabs, put them on all free edges and cut them off when not needed.

Important. Keep one face free of tabs and secure that one last.

3. DELTAHEDRA

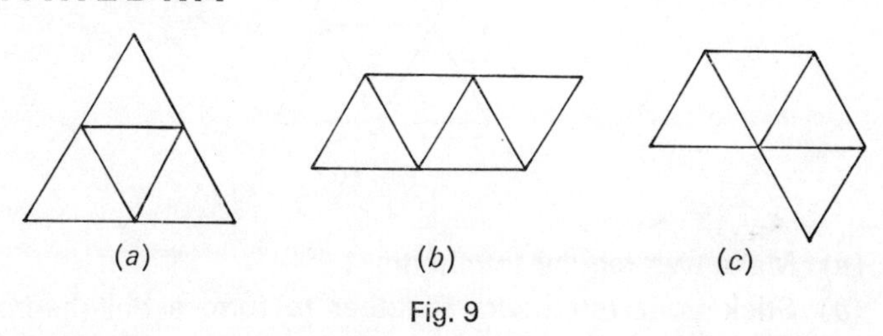

Fig. 9

(*a*) Figure 9 shows the ways in which four equilateral triangles can be joined edge to edge. Which of these shapes form the net of a pyramid with four equilateral triangular faces (including the base)? Check your answer by cutting these shapes from isometric graph paper and folding them.

A pyramid with four faces all triangular is called a *tetrahedron*. If the triangles are equilateral, it is called a *regular* tetrahedron. A regular tetrahedron looks the same no matter on which face you place it.

(*b*) Name another polyhedron which looks the same no matter on which face you place it.

Polyhedra with equilateral triangular faces are called *deltahedra*. A regular tetrahedron is one example of a deltahedron.

(*c*) (i) Make two equal square-based pyramids, each with four equilateral triangular faces.

(ii) Stick the square bases together.

(iii) How many faces has your new polyhedron?

(iv) Is it a deltahedron?

(v) Does it look the same no matter on which face you place it?

(vi) Sketch a net for this new polyhedron.

Your polyhedron has a special name. It is called a regular *octahedron* (see Figure 18).

(*d*) If, for every pair of points of a polyhedron, the line segment joining them lies entirely within the polyhedron, then the polyhedron is said to be *convex*.

Are the following polyhedra convex:

(i) a cube;

(ii) a regular tetrahedron;

(iii) a regular octahedron?

Exercise B

1 Figure 10 shows a net of a regular tetrahedron.

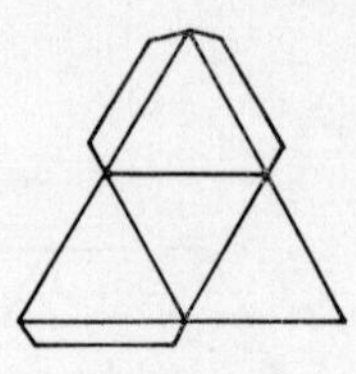

Fig. 10

(*a*) Make two regular tetrahedra.

(*b*) Stick your tetrahedra together to form a deltahedron with six faces.

(*c*) Sketch a net of this new deltahedron. Your net should contain only six equilateral triangles.

3. DELTAHEDRA

Since their faces are all equilateral triangles, deltahedra are the easiest polyhedra to make. Much of the work in Exercises B and C is suitable for group activity and it is not intended that every pupil should make every model mentioned. Some pupils will be fired with great enthusiasm, make a large number of models in their own time and ask for further work. The more able can be referred to *Mathematical Models*, by Cundy and Rollett; others with lower reading ability can be supplied with work cards drawn from this and other books. Some suggestions for further work appear at the end of this chapter.

(*a*) The shapes in Figure 9(*a*) and 9(*b*) are nets for the regular tetrahedron.

(*b*) The cube is the only other regular polyhedron which has been met at this stage.

(*c*) (iii) 8. (iv) Yes. (v) Yes. (vi) There are eleven possible nets for a regular octahedron, excluding mirror images (see Figure I). It is not accidental that there are eleven possible nets for both the cube and the octahedron: they are duals. See *Mathematical Models*, p. 78 (2nd edition).

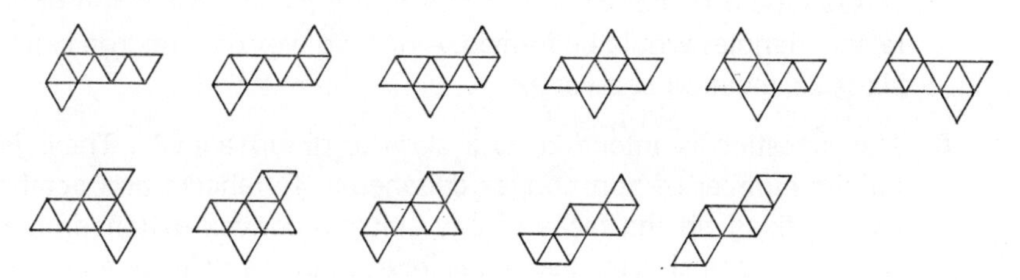

Fig. I. The eleven nets for a regular octahedron.

(*d*) (i) Yes. (ii) Yes. (iii) Yes. The definition of a convex polyhedron will be more fully appreciated when some non-convex polyhedra have been constructed. The similarity between this definition and that for a convex polygon in Chapter 7 can then be discussed.

Exercise B

1 (*c*) Two possible nets for a triangular bi-pyramid are shown in Figure J.

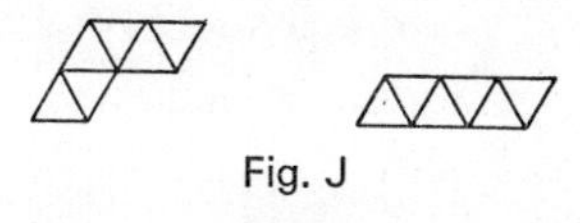

Fig. J

2 See Figure I. Figure 11 (*a*) shows a net for a regular octahedron. There are 66 shapes formed from eight equilateral triangles joined edge to edge. For a pattern of 8 triangles to be a net it must contain at least two instances of 4 triangles at a point. (They can overlap.) If a pattern has 5 or 6 triangles at a point or more than 6 triangles in a line then it cannot be a net for an octahedron.

3 One possible net is shown in Figure K. Note the two sets of five triangles at a point. This polyhedron cannot be formed by fitting five regular tetrahedra together, as the dihedral angle (i.e. the angle between the faces) of the tetrahedron is 70° 32′, slightly less than $\frac{1}{5}$ of 360°.

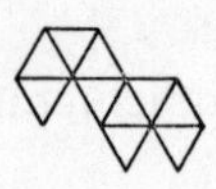

Fig. K

4 The icosahedron is one of the five regular convex polyhedra.

5 For a convex deltahedron, the answer is, 'No'. The faces round the vertex would be coplanar and part of the plane tessellation of equilateral triangles would be formed. A non-convex deltahedron can have six faces meeting at a vertex.

6 This question is intended as a class or group activity. There is an infinite number of non-convex deltahedra. Tetrahedra and octahedra can be fixed on the faces of any suitable convex deltahedron with sometimes amusing results. Some interesting examples of non-convex deltahedra are described in *Mathematical Models,* p. 143.

2 Draw as many shapes as you can formed by eight equilateral triangles joined edge to edge. Figure 11 shows two examples. How many of the shapes which you have drawn are nets for an octahedron? Only one of those in Figure 11 is such a net. Which one?

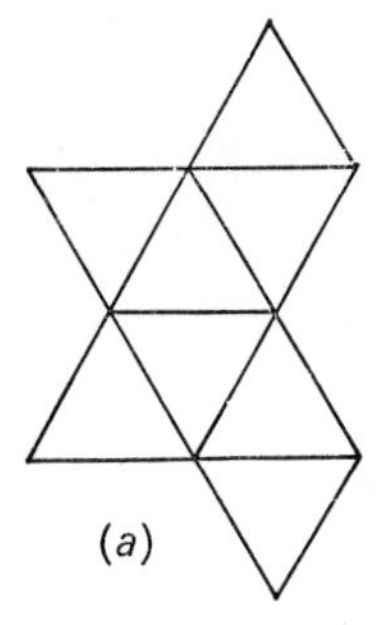

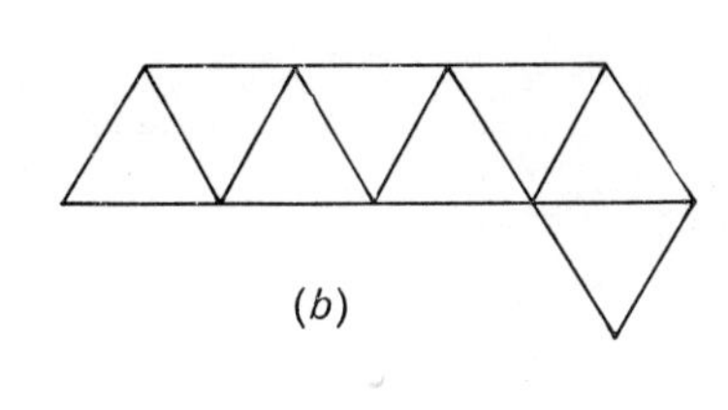

Fig. 11

3 Sketch a net for a ten-faced deltahedron formed by two pyramids with pentagonal bases, joined base to base.

4 One important deltahedron is the icosahedron, which has twenty faces. Figure 12 shows one possible net. Use this net to construct an icosahedron.

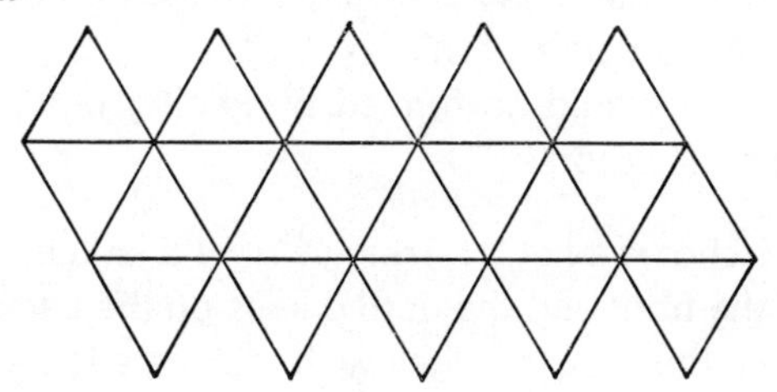

Fig. 12

5 Can a deltahedron have six faces meeting at a vertex? If you think so, make one. If you do not think so, give your reason.

You will need pipe cleaners cut into four pieces and drinking straws cut into two equal parts for Questions 6–8. (Figure 13 shows a neat way of joining three straws.)

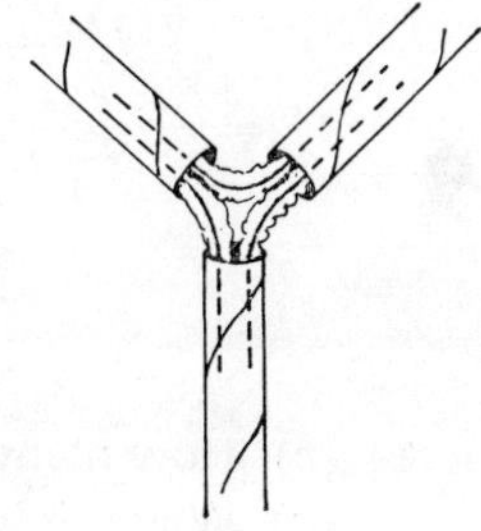

Fig. 13

6 All the deltahedra we have made so far are convex. Use straws and pipe cleaners to make skeleton models of some deltahedra which are not convex. For example, one can be formed by taking a tetrahedron and sticking an equal tetrahedron on to each of its four faces.

7 You have already made convex deltahedra with four, six, eight, ten and twenty faces. There are three more with twelve, fourteen and sixteen faces respectively. Make skeleton models of these three deltahedra. If you have difficulty, first make the models from card using the nets in Figure 14.

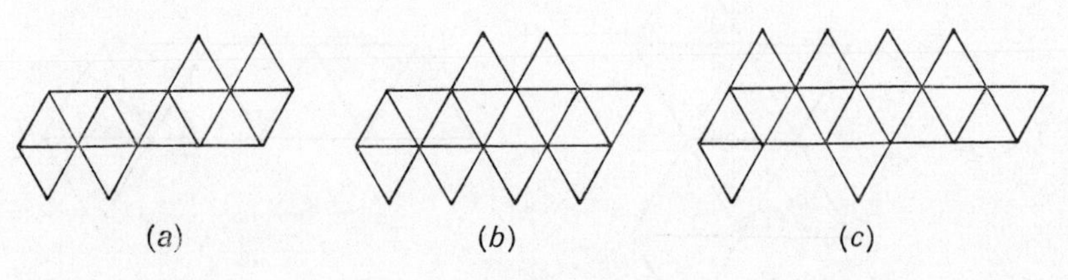

Fig. 14

8 You have now made several skeleton deltahedra. The shape of the triangular faces cannot be changed without bending the straws. Make a skeleton cube. Can the shape of the square faces be easily changed? Which of your models are rigid?

9 Take an octahedron and stick a tetrahedron on to each face. You will get a 'stellated octahedron'.

What solid would be formed by joining the 8 outer vertices of a stellated octahedron?

10 Figure 15 shows a net for a ring of 10 regular tetrahedra. Score the full lines on the front and the dotted lines on the back. Fold the net carefully. Stick the tabs to edges with the same letter (that is, stick tab *a* to edge *a*, and so on). The ring can be turned round and round and is therefore called a rotatable ring of tetrahedra.

You can make a ring with more tetrahedra by using a longer net.

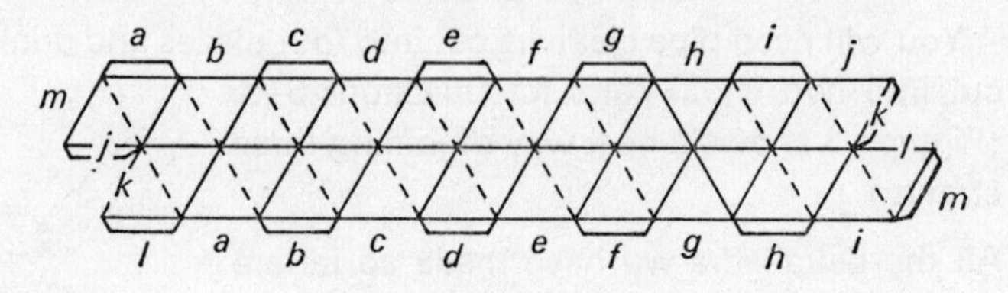

Fig. 15

11 (*a*) How many faces of a polyhedron meet at an edge?

(*b*) Why is it impossible to make a deltahedron with exactly 5 faces?

(*c*) Is it possible to make a deltahedron with exactly 9 faces?

7 There are only eight convex deltahedra.

8 This question is again intended to demonstrate the rigidity of triangular networks.

9 A cube. A stellated octahedron will probably already have been produced by some member or members of the class when answering Question 6 and can be used for answering this question.

10 Best results are obtained with an even number of tetrahedra. If the number is a multiple of 4 then join m to k, k to m, j to l, l to j; the rest as in Figure 15.

For an odd number of tetrahedra either method of joining can be used.

11 (*a*) Two.

(*b*) All the faces of a deltahedron are triangular, so all deltahedra must have an even number of faces. If F is the number of faces and E the number of edges, then $E = \frac{3}{2}F$. For example:

tetrahedron, $F=4$, $E=6$;

octahedron, $F=8$, $E=12$;

icosahedron, $F=20$, $E=30$.

(*c*) No.

4. THE DODECAHEDRON

4.1

(*a*) Tetrahedron, octahedron and icosahedron.

4.2

The five convex regular polyhedra are shown in Figure 18. There are also four non-convex regular polyhedra. It is interesting that these were unknown to the Greeks. The two stellated dodecahedra referred to in the pupil's text were found by Kepler (b. 1571). At first glance, pupils may think that each of these dodecahedra has sixty triangular faces. In fact, each has twelve faces which are regular pentagrams since there are twelve groups of five coplanar triangles. It is for this reason that they are called dodecahedra. One pleasing method of painting these solids is to use a different colour for each of the twelve faces. The other two non-convex regular polyhedra, the great dodecahedron and the great icosahedron, were found by Poinsot (b. 1777). Their nets are given in *Mathematical Models*. Because of the re-entrant angles, these are comparatively difficult to make, though the authors have seen many models successfully completed by pupils aged eleven.

4. THE DODECAHEDRON

4.1 Convex regular polyhedra

There are only five regular polyhedra that are convex. They are the ones with 4, 6, 8, 12 and 20 faces. The one with 6 faces is the cube.

Those with 4, 8 and 20 faces are regular deltahedra. If you did Questions 1, 2 and 4 of Exercise B, you will have already made them. What are their names?

The fifth regular convex polyhedron has 12 faces and is called a *dodecahedron*. Its faces are regular pentagons. The net for this is best constructed in two halves. Draw a large regular pentagon by first drawing a circle. Draw all the diagonals and you will find a smaller pentagon inside the first one. Draw the diagonals of the smaller pentagon making them long enough to meet the sides of the larger one (see Figure 16).

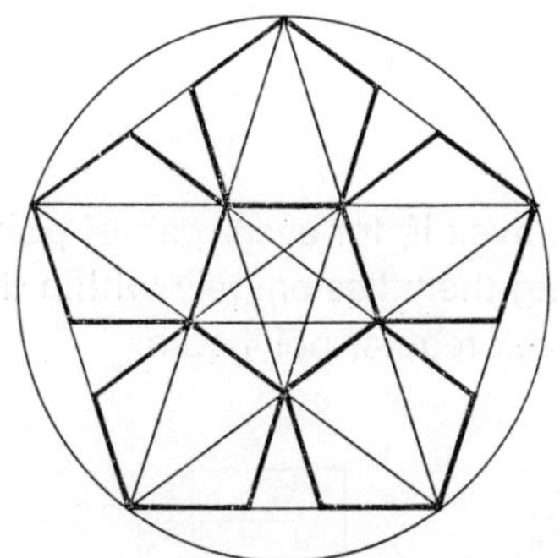

Fig. 16

The thick lines in Figure 16 form half the net of a dodecahedron.

Construct a regular dodecahedron. Do not forget to leave one face free of tabs and secure it last.

4.2 Non-convex regular polyhedra

Two non-convex regular polyhedra are made by adding pyramids to the faces of (*a*) a dodecahedron, (*b*) an icosahedron. The nets of the pyramids are shown in Figure 17. You should find out for yourself how many pyramids are needed in each case and also the correct size to make them. The resulting polyhedra are usually called (*a*) a small stellated dodecahedron; (*b*) a great stellated dodecahedron.

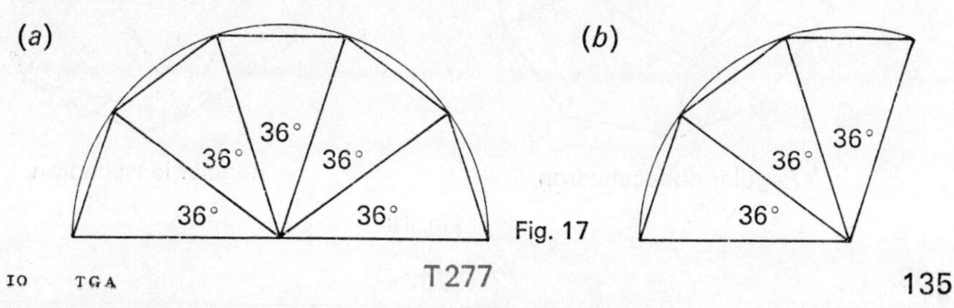

Fig. 17

Exercise C

Class projects

1 Make (*a*) a small stellated dodecahedron; (*b*) a great stellated dodecahedron. Why do you think they are called dodecahedra?

2 Make a list of all the polyhedra you have made. Count the number of faces, vertices and edges for each polyhedron. Enter your result in a table, like the following.

Description of polyhedron	Number of faces (F)	Number of vertices (V)	Number of edges (E)
Cube	6	8	12
Octahedron	8		

Look for patterns in your table and comment on those that you find.

SUMMARY

A polyhedron is convex if, for every pair of points of the polyhedron, the line segment joining them lies entirely within the polyhedron.

There are 5 convex regular polyhedra.

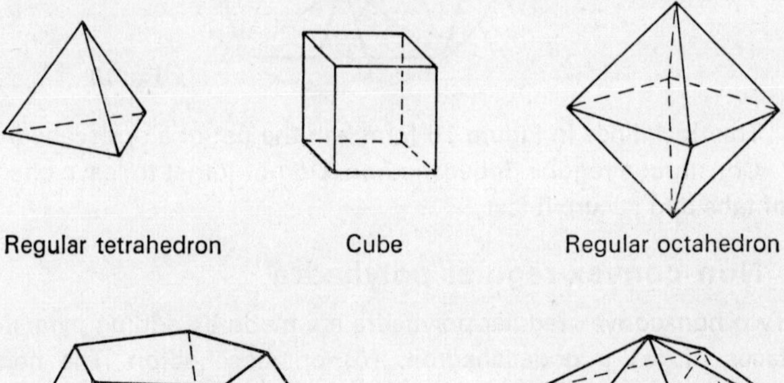

Regular tetrahedron Cube Regular octahedron

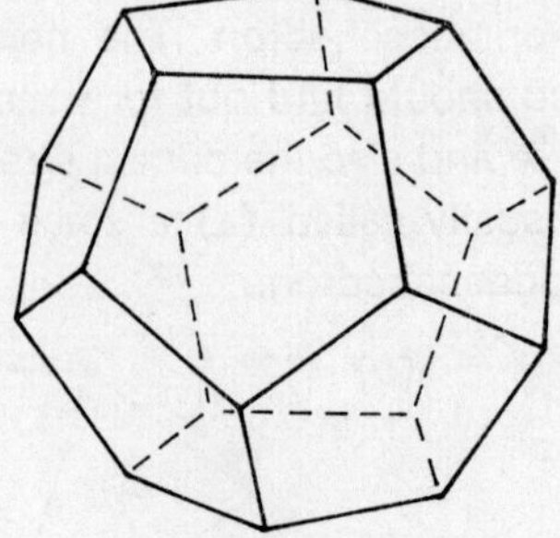

Regular dodecahedron

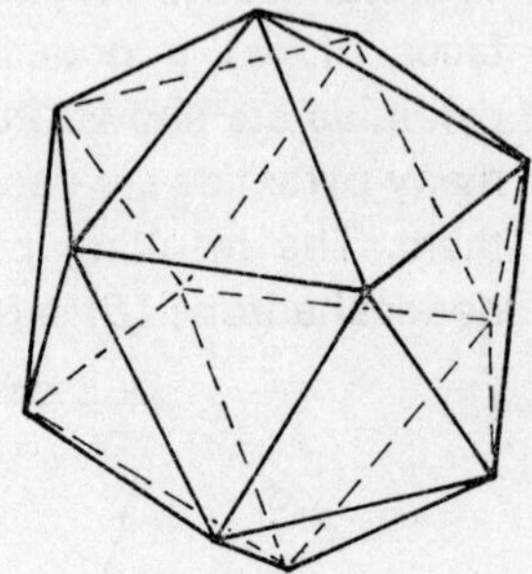

Regular icosahedron

Fig. 18

Exercise C

1 Each has twelve faces. See Section 4.2 above.

2

Description of polyhedron	Number of faces (F)	Number of vertices (V)	Number of edges (E)
Cube	6	8	12
Octahedron	8	6	12
Tetrahedron	4	4	6
Dodecahedron	12	20	30
Icosahedron	20	12	30
Triangular prism	5	6	9
Hexagonal prism	8	12	18
Square pyramid	5	5	8
Truncated square pyramid	6	8	12
6-faced deltahedron	6	5	9
10-faced deltahedron	10	7	15
12-faced deltahedron	12	8	18
14-faced deltahedron	14	9	21
16-faced deltahedron	16	10	24
Small stellated dodecahedron	12	12	30
Great stellated dodecahedron	12	20	30

$F+V-E=2$ for all the polyhedra mentioned except the small stellated dodecahedron. Some pupils will notice the duality of the cube and octahedron and that of the icosahedron and dodecahedron. Others will see a connection between the cube and the truncated square pyramid.

The Euler formula fails to hold in the case of the small stellated dodecahedron because we have chosen to adopt a rather artificial definition of faces, vertices and edges. In one sense, the pupils who at first glance counted too many faces were right! The terminology adopted here is that of *Mathematical Models*, which seeks to preserve the idea of regularity by insisting that the vertices shall lie in certain planes. The Euler formula, on the other hand, really belongs to the subject called algebraic topology and in this context a polyhedron is a system of polygons arranged in such a way that:

(*a*) exactly two polygons meet at an angle at every edge;

(*b*) it is possible to get from any polygon of the system to any other polygon by a path which crosses edges of the polyhedron.

A *simple polyhedron* is one which can be deformed continuously into a sphere. Euler's formula applies to simple polyhedra but it is easy to verify that this is not so in the case of the small stellated dodecahedron. See *Geometry and the Imagination*, by Hilbert and Cohn-Vossen, for further details.

WORK CARDS

The following cards contain suggestions for further work on polyhedra. It is not expected that they will be used in their present form, since the content and language of work cards should depend on the interests and ability of the pupils for whom they are designed. For this reason, answers have not been provided.

For 4 pupils *No.* 1 *Shapes from cubes*

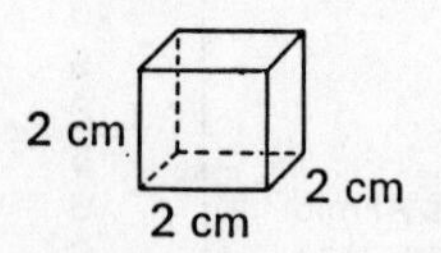

The diagram shows a cube of side 2 cm.

How many different shapes can you make with:

(i) two of these cubes;

(ii) three of these cubes;

(iii) four of these cubes?

Illustrate your answer:

either by constructing models,

or by drawing large clear diagrams.

For 2 pupils *No. 2* *The tetrahedron*

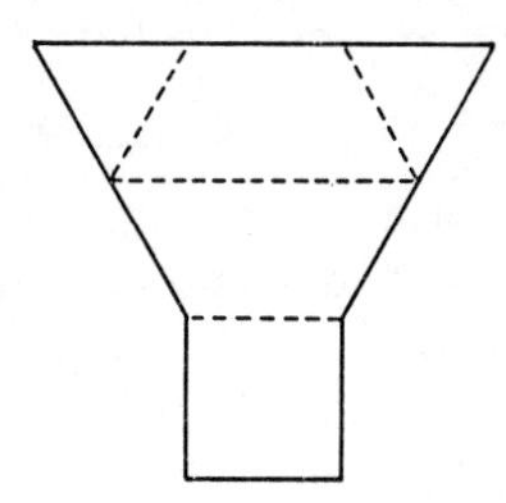

Trace this net and prick through the corners on to heavy cartridge paper. Cut it out and fold along the dotted lines. Either put tabs on and stick it together or secure with sellotape.

Can you use your two polyhedra to make a tetrahedron?

For 2 pupils *No. 3* *The square pyramid*

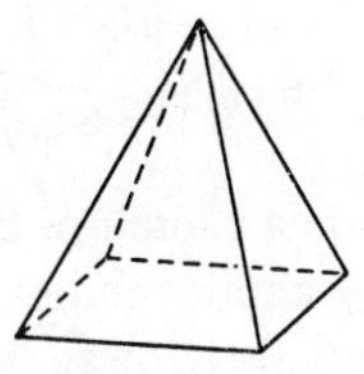

The diagram shows a square pyramid with four congruent faces.

How many faces has this pyramid? How many vertices? How many edges?

Describe the shape of the congruent faces. Are they isosceles triangles? Are they equilateral? Could they be? Are they right-angled? Could they be?

Make a skeleton pyramid from straws and pipe-cleaners. How many straws do you need? Must they all be the same length? Could they be?

Is your pyramid rigid?

Find out anything else that you can about square pyramids.

T

Polyhedra

2 or 4 pupils — No. 4 — *The faces of polyhedra*

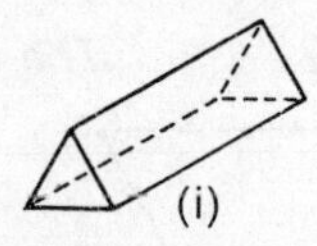

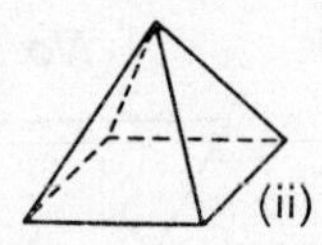

The five faces of a pentahedron are either (i) two triangles and three quadrilaterals; or (ii) four triangles and one quadrilateral.

Find out all you can about the six faces of a hexahedron. Can all the faces be triangles? Can they all be quadrilaterals? Is it possible to have some triangles and some quadrilaterals? Are they any other possibilities?

Display your results in a table and make the polyhedra. (You may use straws and pipe-cleaners or card and glue.)

For 2 pupils — *No.* 5 — *Cutting a cube*

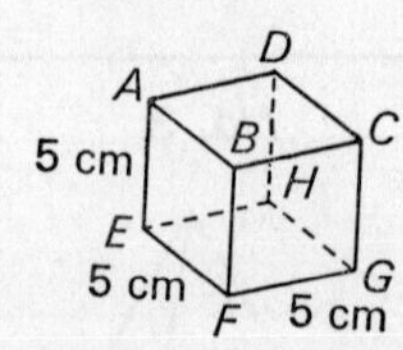

The diagram shows a sketch of a solid cube.

What can you say about the lengths of *BD*, *BE* and *DE*? Suppose you cut the cube so that the corner *AEBD* is cut off. Describe the shape of the solid *AEBD*.

Describe the shape of each of its faces.

Draw a net for the solid *AEBD*.

Book list

Mathematical Models (2nd edition). Cundy and Rollett.
Mathematical Snapshots. Steinhaus.
Experiments in Mathematics, Stage 1, 2, 3. Pearcy and Lewis.
Polyhedra. Lewis. Longmans, 1968.
Polyominoes. Golomb.
Mathematical Puzzles and Diversions. Gardner.
More Mathematical Puzzles and Diversions. Gardner.
Introduction to Geometry. Coxeter.
Geometry and Imagination. Hilbert and Cohn-Vossen.

A pyramid is identified by referring to the shape of its base. Figure 19 (*a*) shows a hexagonal pyramid.

A prism is identified by referring to the shape of its 'ends'. Figure 19 (*b*) shows a triangular prism.

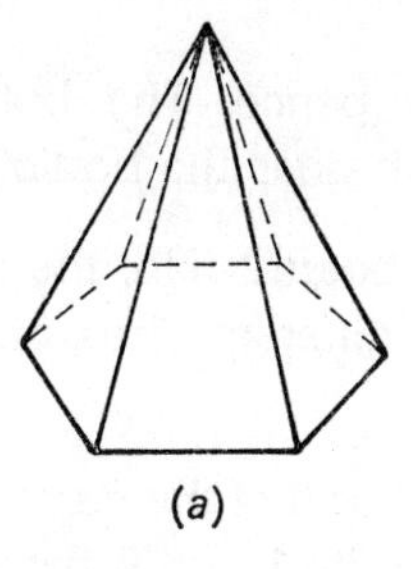

(*a*)

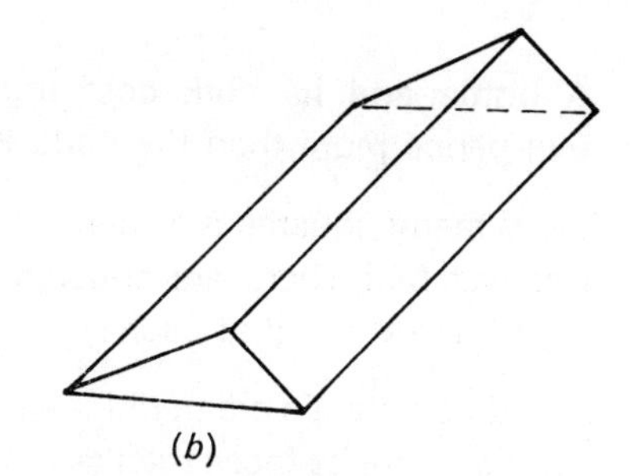

(*b*)

Fig. 19

Puzzle corner

1 A bottle and its cork cost together six pence. The bottle costs five pence more than the cork. How much does the bottle cost?

2 How many squares are there on a chess board? (No, the answer is not just 64! There are squares of eight different sizes on a chess board. How many of each?)

3 The dots on a die are traditionally arranged so that the number of dots on the opposite faces total seven. Keeping the 1 and 6 fixed, in how many ways can the 2 and 5 and 3 and 4 be arranged round them?

4 A frog is at the bottom of a 10 m well. Each hour he climbs up 1 m and then slips back 0·5 m. How many hours does it take him to get out?

5 A park gardener took a great pride in his flower beds. In particular, he liked to put the plants in straight lines. Imagine his delight when one day he discovered that he had planted 9 geraniums so that there were 10 lines with 3 plants in each. Make a drawing to show how 9 geraniums can be planted with just 3 plants in each line making (*a*) 8 lines (see Prelude), (*b*) 9 lines, (*c*) 10 lines.

6 What comes next?

(*a*) 5, 7, 11, 19, . . . ;
(*b*) 2, 6, 12, 20, . . . ;
(*c*) 0, 1, 1, 2, 3, 5, . . . ;
(*d*) 2, 3, 5, 7, 11, ...

7 A boy cycles 3 km to his school each morning. He leaves home at a time which allows him to average 15 km/h and arrive for roll call. One morning the traffic was heavy and he only managed to average 10 km/h for the first two kilometres. What speed must he average for the last kilometre to be in time for roll call?

8 A clock strikes six in 5 seconds. How long does it take to strike twelve?

9 In the following addition each letter stands for a number. What are the other numbers given that E stands for 1 and U stands for 9?

```
 THREE
 THREE
  FOUR
------
ELEVEN
```

T

Puzzle corner

These questions are rather harder than those in the rest of the book and it should be stated that the authors do not think they are suitable in a context where only limited time is available. They are the sort of questions that some people enjoy puzzling over but they have no method which, at this stage, can be isolated, set in context and generalized.

1 The bottle costs $5\frac{1}{2}$p, and the cork $\frac{1}{2}$p.

2 204 squares.

This is a useful exercise in pattern spotting.
Number of 'one unit' squares $= 8^2$.
Number of 'two unit' squares $= 7^2$.
....
Total number of squares $= 8^2+7^2+6^2+5^2+4^2+3^2+2^2+1^2$.

3 8 ways. The arrangements as shown in Figure A, together with the four obtained when the letters are interchanged.

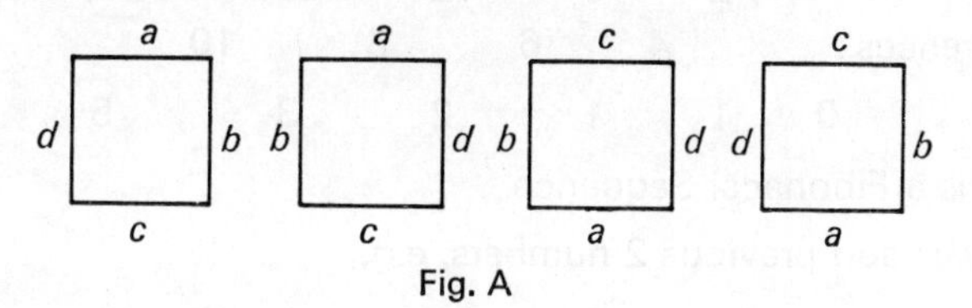

Fig. A

4 19 hours.
After the 18th hour the frog has climbed 9 m.
During the 19th hour it climbs 1 m and reaches the top.

5 See Figure B. This is a special case of the theorem in projective geometry known as Pappus's Theorem.

A similar problem is to plant 10 geraniums in 10 rows of 3. The solution is given by the configuration of Desargues's Theorem. It may be worth mentioning at this stage the existence of a geometry without measurement: Pappus's and Desargues's Theorems, and the harmonic range construction are useful examples.

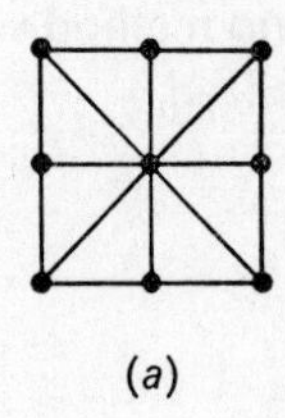
(*a*)

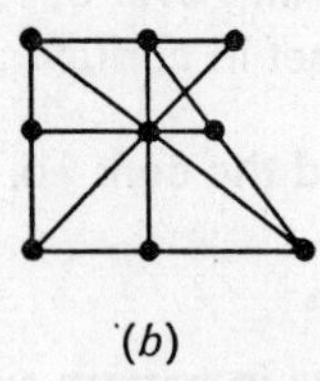
(*b*)

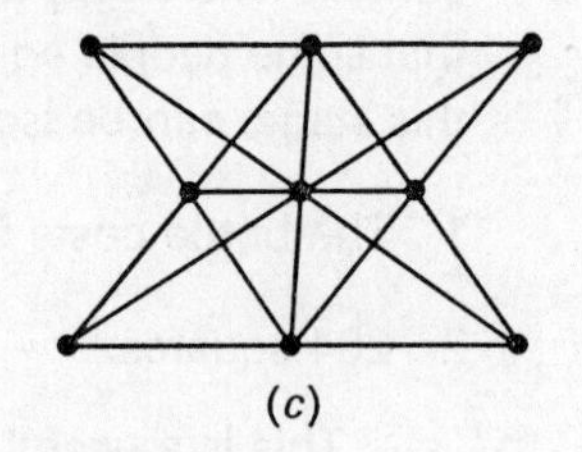
(*c*)

Fig. B

References: for a detailed study, any book on projective geometry; for the basic ideas, *What is Mathematics?*, by Courant and Robbins.

6 (*a*) 5 7 11 19 35;

differences: 2 4 8 16

(*b*) 2 6 12 20 30;

differences: 4 6 8 10

(*c*) 0 1 1 2 3 5 8.

This is a Fibonacci Sequence.

Rule: add previous 2 numbers, e.g.

$$0+1=1,$$

$$1+1=2,$$

$$1+2=3, \text{ etc.}$$

This could be followed up with examples of Fibonacci Sequences, e.g. ancestry of a male bee, breeding rabbits, phyllotaxis. Book 2 contains some work on these and other aspects of Fibonacci Sequences.

References: *The Language of Mathematics*, by Land; *Riddles in Mathematics*, by Northrop.

(*d*) 2 3 5 7 11 13. Prime numbers!

7 He cannot arrive in time.

8 11 s (assuming negligible striking times).
There are 5 pauses in striking six o'clock: hence each pause takes 1 s. There are 11 pauses in striking twelve o'clock.

9 There is not a unique solution.

74 611	84 011
74 611	84 011
2 096	3 590
151 318	171 612

Solve by a combination of logic and trial-and-error.

10 3 weighings.

Split the balls into 3 sets of 9, *A*, *B* and *C*.

Weigh *A* against *B*. This determines the heavy set (if *A* and *B* balance, the heavy ball is in *C*).

Split the heavy set of 9 into 3 sets of 3.

Repeat the procedure.

One weighing then determines the heavy ball from the heavy set of 3.

11 The 6 arrangements are shown in Figure C.

12

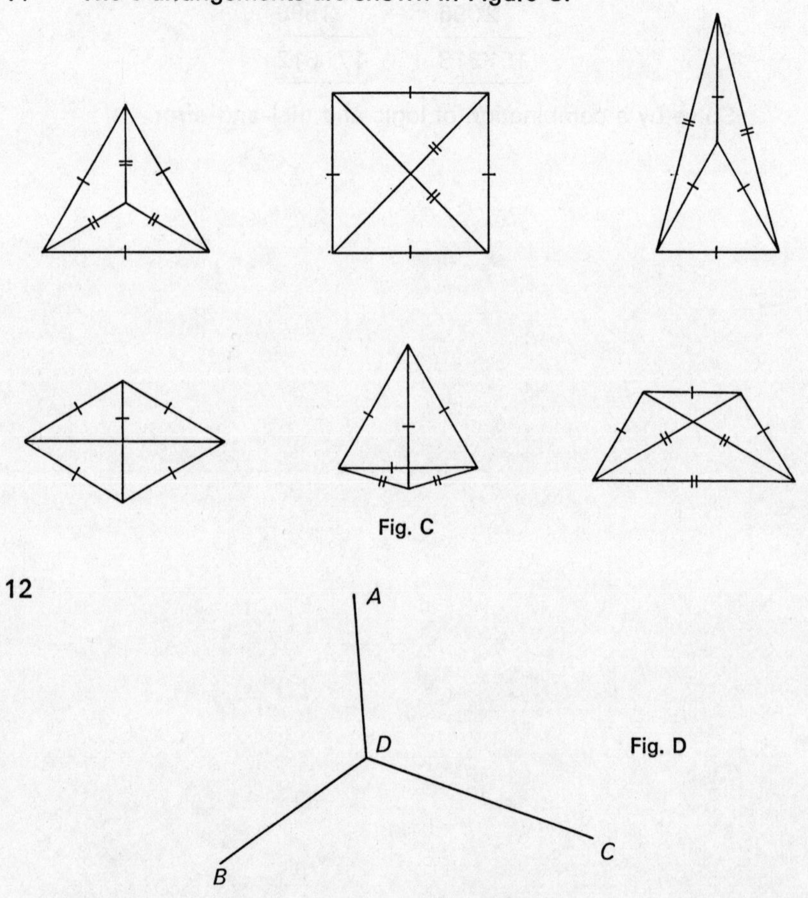

Fig. C

Fig. D

D (see Figure D) is the point at which *AB*, *BC*, *CA* all subtend an angle of 120°. If one angle, say *A*, of triangle *ABC* had been equal to or greater than 120°, then *D* would coincide with *A*. For a full discussion see Steiner's Problem in *What is Mathematics?*, by Courant and Robbins.

10 It is known that a faulty billiard ball has been mixed up with 26 good balls. The balls all look alike but the faulty one is heavier than the others. How can the faulty ball be detected by weighing balls against one another on a pair of scales? What is the smallest number of weighings required?

11 Mark 4 points on paper so that there are only 2 different distances between them. One arrangement is shown in Figure 1.

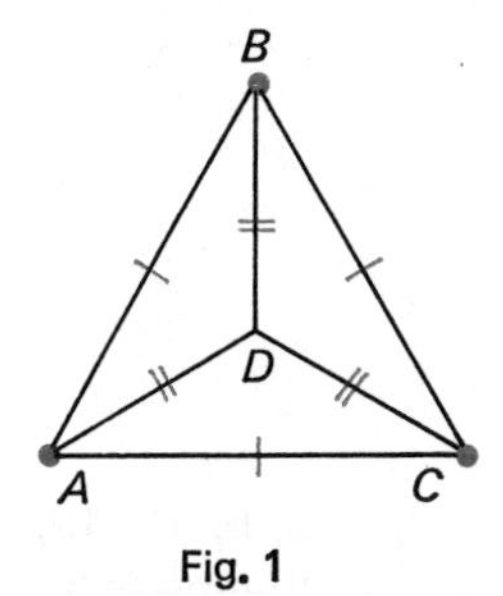

$AB = BC = CA,$
$DA = DB = DC.$

Fig. 1

There are 6 different possible arrangements. Find two others.

12 A civil engineer had to design a road system connecting 3 towns *A*, *B* and *C* (see Figure 2).

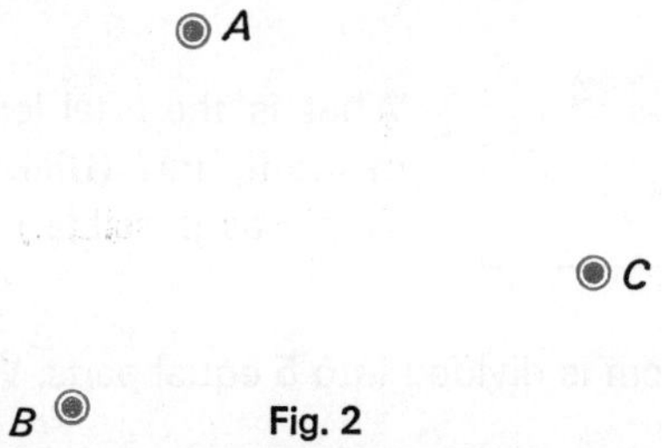

$AB = 5$ km,
$BC = 7$ km,
$CA = 6$ km.

Fig. 2

Naturally the county council wanted to keep the roads as short as possible to reduce the cost. Draw accurate maps of the towns and, by trial and error, find as short a road system as you can. State its length.

Revision exercises

Quick quiz, no. 3

1 Which of these figures are *regular* polygons?

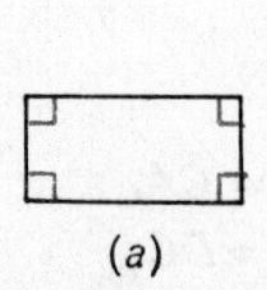

(*a*)

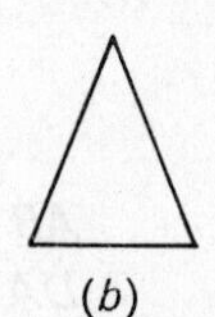

(*b*)

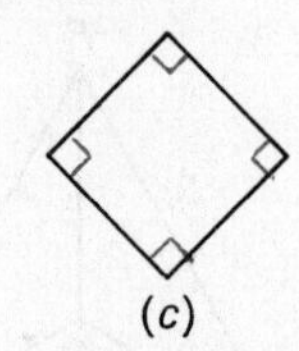

(*c*)

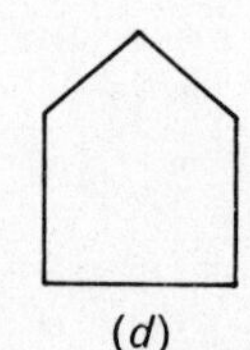

(*d*)

2 $\frac{1}{9}+\frac{2}{9}+\frac{4}{9}=?$

3 If $A=\{\text{factors of } 60\}=\{1, 2, 3, 4, 5, 6, 10, 12, 15, 20, 30, 60\}$ and $B=\{\textit{prime}\text{ factors of } 60\}$, what are the members of B?

4 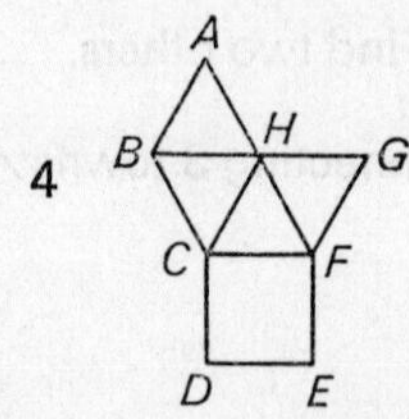

This is a net for a polyhedron. When it is 'made up', what points will be brought to E?

5 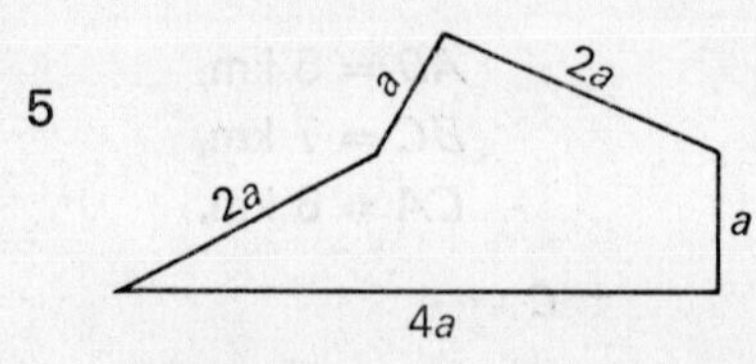

What is the total length of the sides of the figure? (Make your answer as simple as possible.)

6 A line of length 4 cm is divided into 5 equal parts. What is the length of 1 part?

Quick quiz, no. 4

1 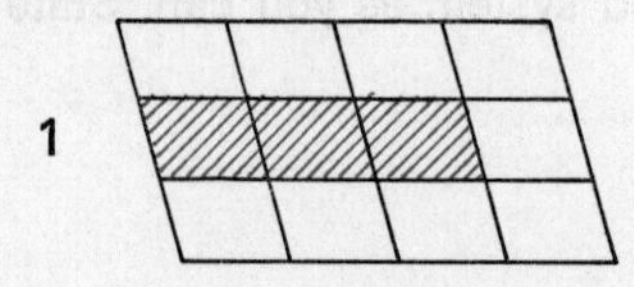

What fraction of this figure is shaded?

2 How many diagonals could be drawn from any one vertex of a 24-sided convex polygon?

3 10 is the fourth triangle number. What is the sixth triangle number?

Revision exercises

Quick quiz, no. 3

1 (c).

2 $\frac{7}{9}$.

3 2, 3, 5.

4 A and G.

5 $10a$.

6 $\frac{4}{5}$ cm.

Quick quiz, no. 4

1 $\frac{3}{12}$ or $\frac{1}{4}$.

2 21.

3 21.

Revision exercises

4 8.

5 $\frac{4}{7}$.

6 True.

Exercise C

1

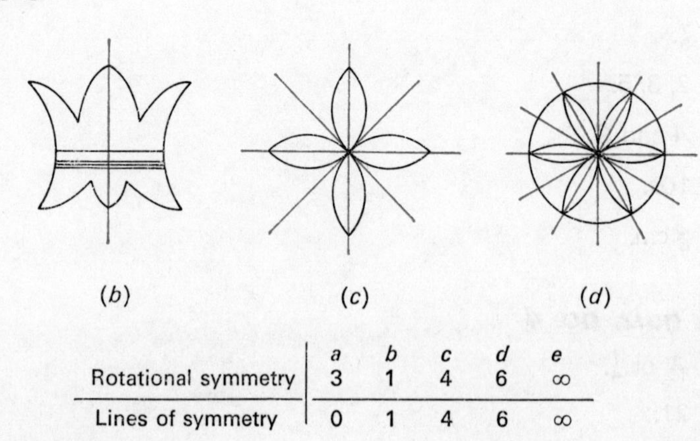

(*b*) (*c*) (*d*)

	a	*b*	*c*	*d*	*e*
Rotational symmetry	3	1	4	6	∞
Lines of symmetry	0	1	4	6	∞

2 (*a*) False;

(*b*) true;

(*c*) false;

(*d*) false;

(*e*) false.

3 No. The sum of the remaining four angles must be 720°. If they were equal then each would be 180°: this is impossible so one at least must be greater than 180° and the figure must be non-convex (re-entrant).

4 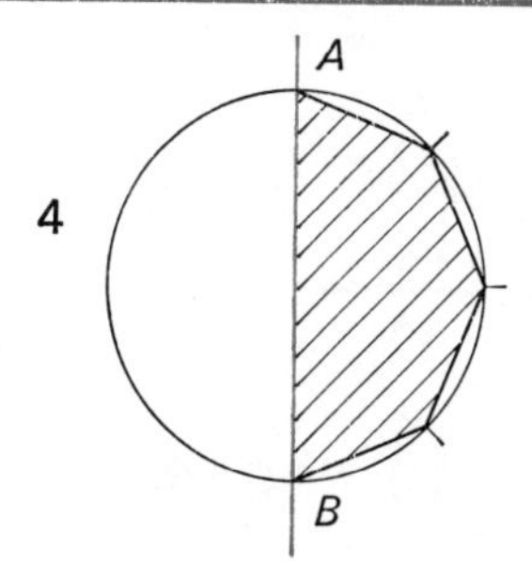

AB is a line of symmetry of a regular polygon, part of which is shown. How many lines of symmetry has the completed figure?

5 17 identical glasses are filled from 3 bottles of cider. Each bottle contains seven of these glassfuls. What fraction of a bottle of cider remains?

6 $1048 \div 8 = (1000 \div 8) + (48 \div 8)$. True or false?

Exercise C

1 Trace each figure. Put in any lines of symmetry.
What rotational symmetry does it have about the marked point?

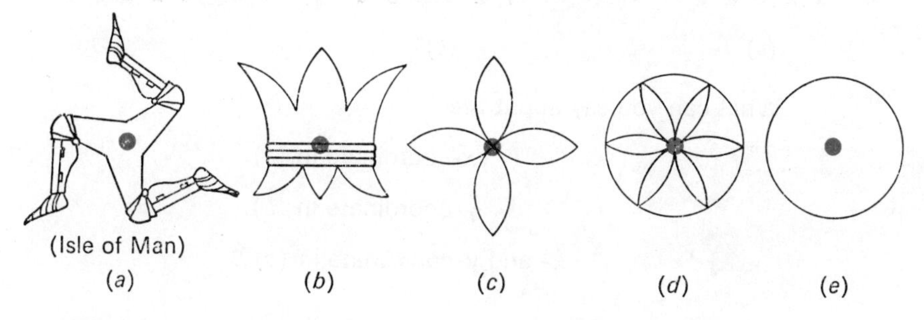

(Isle of Man)
(*a*) (*b*) (*c*) (*d*) (*e*)

2 Which of the following statements are true and which are false?

(*a*) No prime number is even;
(*b*) the square of an odd number is always an odd number;
(*c*) two square numbers added together never make another square number;
(*d*) two prime numbers multiplied together sometimes make another prime number;
(*e*) apart from 1, no triangle number is also a square number.

3 Draw an eight-sided figure with:

1 line of symmetry,
4 right-angles,
3 adjacent sides equal and the other 5 equal to one another.

Is your figure convex? If not, do you think it possible to draw a figure with these properties that is convex?

4 If $P = \{\text{factors of } 54\}$,

$Q = \{\text{prime factors of } 42\}$,

$R = \{\text{multiples of 3 that are less than } 30\}$,

list the members of the sets P, Q and R.
List the members of the sets:

(*a*) $P \cap Q$; (*b*) $P \cap R$; (*c*) $P \cap Q \cap R$.

5

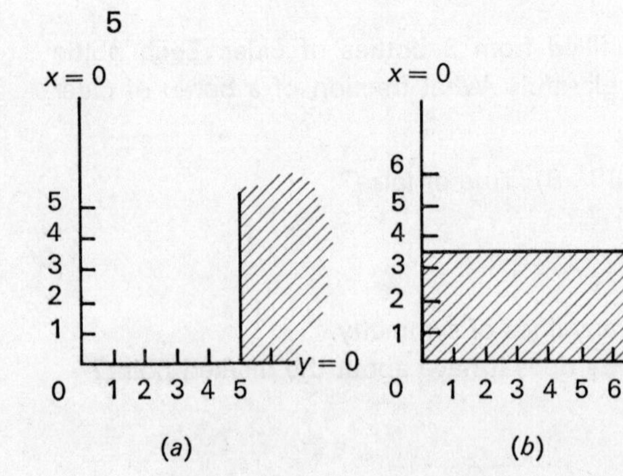

(*a*) (*b*)

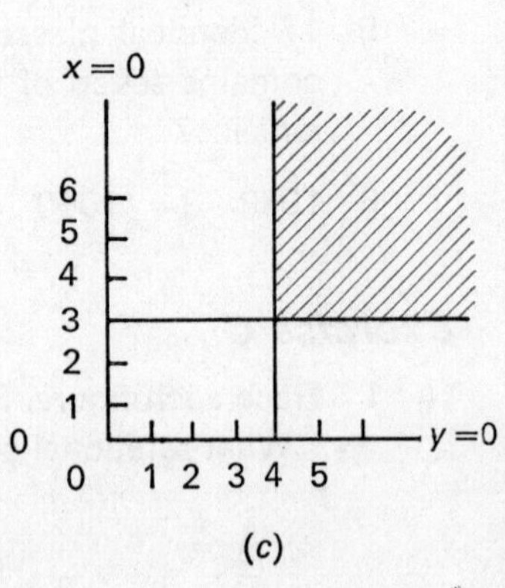

(*c*)

What can you say about the

x-coordinate in (*a*),

y-coordinate in (*b*),

x- and *y*-coordinate in (*c*)?

Exercise D

1 Using squared paper, draw accurately the net for the prism shown:

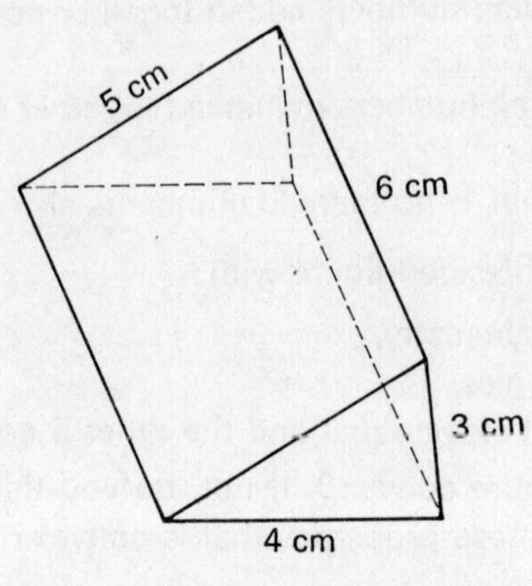

4 $P = \{1, 2, 3, 6, 9, 18, 27, 54\}$,
$Q = \{2, 3, 7\}$,
$R = \{3, 6, 9, 12, 15, 18, 21, 24, 27\}$.
(*a*) $P \cap Q = \{2, 3\}$;
(*b*) $P \cap R = \{3, 6, 9, 18, 27\}$;
(*c*) $P \cap Q \cap R = \{3\}$.

5 $x > 5$, $0 < y < 3\frac{1}{2}$, $x > 4$ and $y > 3$.

Exercise D

1

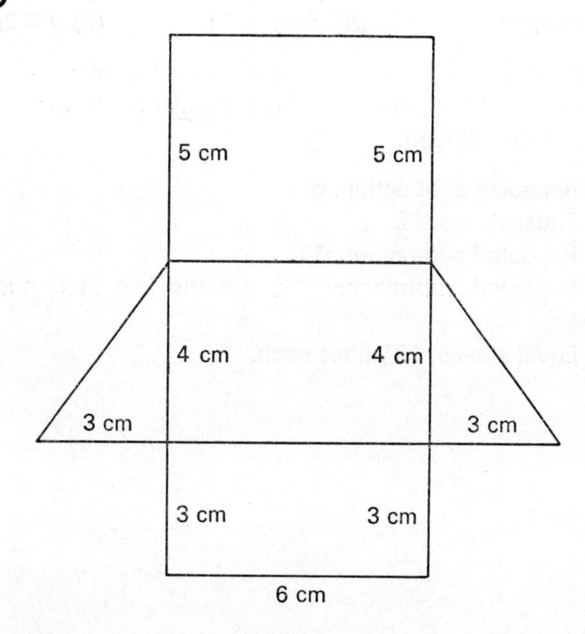

Revision exercises

2 (*a*) 3 and 13, 5 and 11;
(*b*) 3 and 2;
(*c*) 2 and any other prime;
(*d*) $48 = 2^4 \times 3$.

3

Number	18	24	25	30	36	40	49	60	64
Number of distinct rectangular patterns (p)	2	3	1	3	4	3	1	5	3
Number of factors excluding 1 and the number itself (f)	4	6	1	6	7	6	1	10	5

(*a*) $f = 2p$; (*b*) $f = p = 1$; (*c*) $f = 2p - 1$.

4 (*a*) Square; (*b*) parallelogram;
(*c*) equilateral triangle; (*d*) regular hexagon;
(*e*) regular octagon.

5 (*a*) Repeated subtraction. 5.
(*b*) Equal shares. 12.
(*c*) Repeated subtraction. 11.
(*d*) Repeated subtraction. 12, but the last box must hold extra packing.
(*e*) Equal shares. $12\frac{1}{2}$ litres each.

2 (*a*) Write down two pairs of prime numbers whose sum is 16.
(*b*) Write down two prime numbers whose difference is 1.
(*c*) Write down two prime numbers whose sum is an odd number.
(*d*) Factorize 48 into primes.

3 Copy and complete the following table. (Note that we are to count only distinct patterns; the two dot patterns shown are considered to be the same.)

Number	18	24	25	30	36	40	49	60	64
Number (p) of distinct rectangular patterns	2								
Number (f) of factors (excluding 1 and the number itself)	4								

Can you find a connection between 'p' and 'f' for:
(*a*) rectangle numbers (excluding square numbers);
(*b*) odd square numbers;
(*c*) even square numbers?

4 Identify the following polygons:
(*a*) 4 lines of symmetry, 4 sides;
(*b*) no lines of symmetry, but rotational symmetry of order 2 and 4 sides;
(*c*) 3 sides and 3 lines of symmetry;
(*d*) 6 lines of symmetry and 6 sides;
(*e*) 8 sides and rotational symmetry of order 8.

5 Which of the following are examples of division as repeated subtraction and which are examples of division as a splitting into equal parts? Write down the numerical answer to each question.
(*a*) How many people can be given exactly 3 fish fingers out of a packet of 16?
(*b*) How many oranges would each person get if a box of 48 was divided between four people?
(*c*) How many sandwiches can be made from a cut loaf of 22 slices?
(*d*) Some cardboard boxes are made to take 60 tins. How many boxes are required if 700 tins are to be packed?
(*e*) 100 litres of water has to be divided between 8 people. How much should each person get?

Revision exercises

Exercise E

1 Copy and complete the patterns shown in the figure. The dotted lines are lines of symmetry, and the red dots are centres of rotational symmetry.

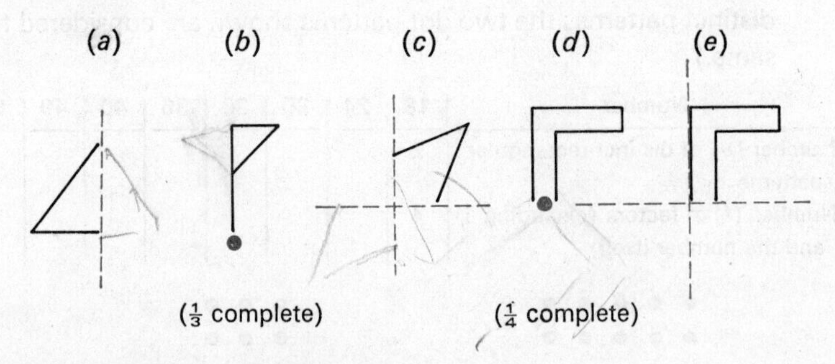

2 If A = {prime factors of 60},

B = {factors of 20 (excluding 1 and 20)},

C = {square numbers less than 75},

write down the members of each set.
What are: (*a*) $A \cap B$; (*b*) $A \cap C$?

3 A man on holiday in Switzerland buys 6 bread rolls at 30 cents each and 3 apple tarts at 50 cents each. How much change does he get from a 5 franc note if there are 100 cents in 1 franc?

4 If the earth revolves about its axis once every 24 hours, through what angle (in degrees) does it turn

(*a*) in 6 hours; (*b*) in 1 hour;
(*c*) in 4 minutes; (*d*) in 2 days?

How long does it take to turn through 20°?

5 A numeral system uses four digits $*$, ∇, ?, $\oplus$ (not necessarily in that order) and

$$* \nabla ? \oplus + ? \nabla * \nabla = \nabla \nabla ? \nabla \nabla.$$

What do the symbols $*$, ∇, ?, $\oplus$, represent? Write down the first twenty numbers in this system.

Exercise E

1

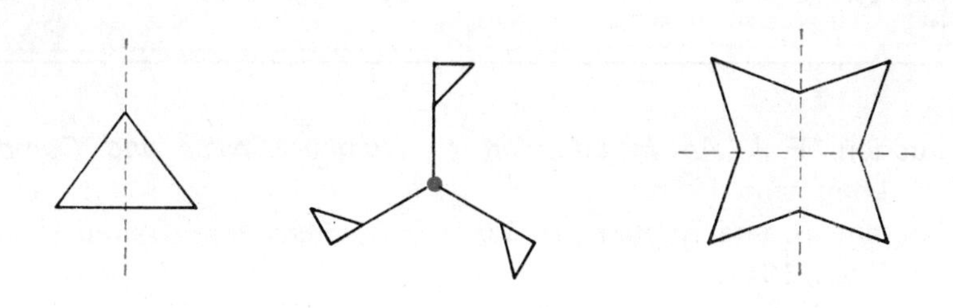

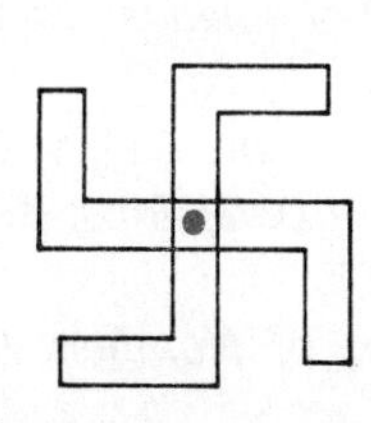

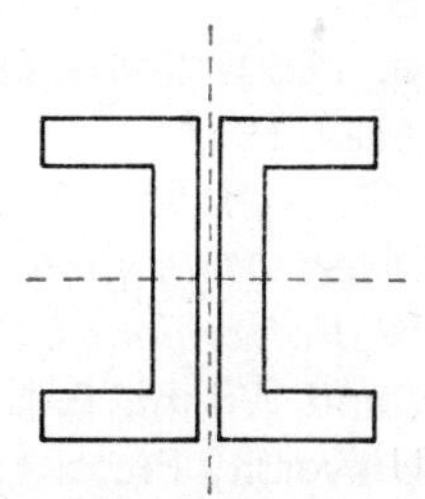

2 $A = \{2, 3, 5\}$,

$B = \{2, 4, 5, 10\}$,

$C = \{1, 4, 9, 16, 25, 36, 49, 64\}$.

(i) $A \cap B = \{2, 5\}$; (ii) $A \cap C =$ the empty set.

3 His change is 1 fr. 70 cents.

4 (*a*) 90°; (*b*) 15°; (*c*) 1°; (*d*) 720°.

1 hour, 20 minutes.

5 Using numbers to base ten, ∗ = 2; ∇ = 1; ? = 3; ⊕ = 0.

∇, ∗, ?

∇⊕, ∇∇, ∇∗, ∇?,

∗⊕, ∗∇, ∗∗, ∗?,

?⊕, ?∇, ?∗, ??,

∇⊕⊕, ∇⊕∇, ∇⊕∗, ∇⊕?,

∇∇⊕, ∇∇∇, ∇∇∗, ∇∇?.

T

Bibliography

Budden, F. J. *An Introduction to Number Scales and Computers.* Longmans, 1965.

Courant, R. and Robbins, H. *What is Mathematics?* Oxford University Press, 1941.

Coxeter, H. S. M. *Introduction to Geometry*. Wiley, 1961.

Cundy, H. M. and Rollett, A. P. *Mathematical Models* (2nd edition). Oxford University Press, 1961.

Fletcher, T. J. (Ed.). *Some Lessons in Mathematics.* Cambridge University Press, 1964.

Gardner, M. *Mathematical Puzzles and Diversions*. Bell, 1961.

Gardner, M. *More Mathematical Puzzles and Diversions*. Bell, 1961.

Golomb, S. W. *Polyominoes*. Scribners, 1965.

Hardy, G. H. and Wright, E. M. *The Theory of Numbers* (4th edition). Oxford University Press, 1960.

Hilbert, D. and Cohn-Vossen. *Geometry and the Imagination*. Chelsea, N.Y., 1952.

Land, F. W. *The Language of Mathematics*. Murray, 1960.

Levi, H. *Elements of Algebra* (4th edition). Chelsea, N.Y., 1961.

Lewis, K. *Further Experiments in Mathematics:* topics, *Geometry without Instruments, Polyhedra*. Longmans, 1968.

Mansfield, D. E. and Thompson, D. *Mathematics, A New Approach*. Chatto and Windus, 1962.

Northrop, E. P. *Riddles in Mathematics*. English University Press, 1945.

Pearcy, J. F. F. and Lewis, K. *Experiments in Mathematics*, Stage 1, 2, 3. Longmans, 1966.

Shanks, D. *Solved and Unsolved Problems in Number Theory*. Spartan Books, 1962.

Steinhaus, H. *Mathematical Snapshots*. Oxford University Press, 1960.

Thurston, H. A. *The Number-System*. Interscience Publishers Inc. N.Y., 1956.

Weyl, H. *Symmetry*. Princeton University Press, 1952.

Index

(*Note: references are to the red T page numbers*)

Index

Index

Index